Essential
Understanding
Series

Developing
Essential Understanding
of
Geometry and Measurement

for Teaching Mathematics *in*
Prekindergarten–Grade 2

E. Paul Goldenberg
Educational Development Center
Newton, Massachusetts

Douglas H. Clements
University of Denver
Denver, Colorado

Barbara J. Dougherty
Volume Editor
University of Missouri
Columbia, Missouri

Rose Mary Zbiek
Series Editor
The Pennsylvania State University
University Park, Pennsylvania

NATIONAL COUNCIL OF
TEACHERS OF MATHEMATICS

Copyright © 2014 by
The National Council of Teachers of Mathematics, Inc.
1906 Association Drive, Reston, VA 20191-1502
(703) 620-9840; (800) 235-7566; www.nctm.org
All rights reserved

Library of Congress Cataloging-in-Publication Data

Goldenberg, E. Paul (Ernest Paul) author
  Developing essential understanding of geometry and measurement for teach-
ing mathematics in prekindergarten-grade 2 / by E. Paul Goldenberg, Educational
Development Center, Newton, Massachusetts and Douglas H. Clements, University
of Denver, Denver, Colorado ; Barbara J. Dougherty, volume editor, University of
Missouri Columbia, Missouri.
        pages cm. — (Essential understanding series)
  Includes bibliographical references.
  ISBN 978-0-87353-665-3
  1. Measurement—Study and teaching (Early childhood) 2. Geometry—Study and
teaching (Early childhood) I. Clements, Douglas H., author.
  QA465.G65 2014
  372.7'6--dc23
                                    2013041285

The National Council of Teachers of Mathematics is the public voice of mathematics education,
supporting teachers to ensure equitable mathematics learning of thehighest quality for all
students through vision, leadership, professional development, and research.

Printed in the United States of America

Contents

Foreword

Teaching mathematics in prekindergarten–grade 12 requires a special understanding of mathematics. Effective teachers of mathematics think about and beyond the content that they teach, seeking explanations and making connections to other topics, both inside and outside mathematics. Students meet curriculum and achievement expectations when they work with teachers who know what mathematics is important for each topic that they teach.

The National Council of Teachers of Mathematics (NCTM) presents the Essential Understanding Series in tandem with a call to focus the school mathematics curriculum in the spirit of *Curriculum Focal Points for Prekindergarten through Grade 8 Mathematics: A Quest for Coherence*, published in 2006, and *Focus in High School Mathematics: Reasoning and Sense Making,* released in 2009. The Essential Understanding books are a resource for individual teachers and groups of colleagues interested in engaging in mathematical thinking to enrich and extend their own knowledge of particular mathematics topics in ways that benefit their work with students. The topic of each book is an area of mathematics that is difficult for students to learn, challenging to teach, and critical for students' success as learners and in their future lives and careers.

Drawing on their experiences as teachers, researchers, and mathematicians, the authors have identified the big ideas that are at the heart of each book's topic. A set of essential understandings— mathematical points that capture the essence of the topic—fleshes out each big idea. Taken collectively, the big ideas and essential understandings give a view of a mathematics that is focused, connected, and useful to teachers. Links to topics that students encounter earlier and later in school mathematics and to instruction and assessment practices illustrate the relevance and importance of a teacher's essential understanding of mathematics.

On behalf of the Board of Directors, I offer sincere thanks and appreciation to everyone who has helped to make this series possible. I extend special thanks to Rose Mary Zbiek for her leadership as series editor. I join the Essential Understanding project team in welcoming you to these books and in wishing you many years of continued enjoyment of learning and teaching mathematics.

Henry Kepner
President, 2008–2010
National Council of Teachers of Mathematics

Preface

From prekindergarten through grade 12, the school mathematics curriculum includes important topics that are pivotal in students' development. Students who understand these ideas cross smoothly into new mathematical terrain and continue moving forward with assurance.

However, many of these topics have traditionally been challenging to teach as well as learn, and they often prove to be barriers rather than gateways to students' progress. Students who fail to get a solid grounding in them frequently lose momentum and struggle in subsequent work in mathematics and related disciplines.

The Essential Understanding Series identifies such topics at all levels. Teachers who engage students in these topics play critical roles in students' mathematical achievement. Each volume in the series invites teachers who aim to be not just proficient but outstanding in the classroom—teachers like you—to enrich their understanding of one or more of these topics to ensure students' continued development in mathematics.

How much do you need to know?

To teach these challenging topics effectively, you must draw on a mathematical understanding that is both broad and deep. The challenge is to know considerably more about the topic than you expect your students to know and learn.

Why does your knowledge need to be so extensive? Why must it go above and beyond what you need to teach and your students need to learn? The answer to this question has many parts.

To plan successful learning experiences, you need to understand different models and representations and, in some cases, emerging technologies as you evaluate curriculum materials and create lessons. As you choose and implement learning tasks, you need to know what to emphasize and why those ideas are mathematically important.

While engaging your students in lessons, you must anticipate their perplexities, help them avoid known pitfalls, and recognize and dispel misconceptions. You need to capitalize on unexpected classroom opportunities to make connections among mathematical ideas. If assessment shows that students have not understood the material adequately, you need to know how to address weaknesses that you have identified in their understanding. Your understanding must be sufficiently versatile to allow you to represent the mathematics in different ways to students who don't understand it the first time.

In addition, you need to know where the topic fits in the full span of the mathematics curriculum. You must understand where your students are coming from in their thinking and where they are heading mathematically in the months and years to come.

Accomplishing these tasks in mathematically sound ways is a tall order. A rich understanding of the mathematics supports the varied work of teaching as you guide your students and keep their learning on track.

How can the Essential Understanding Series help?

The Essential Understanding books offer you an opportunity to delve into the mathematics that you teach and reinforce your content knowledge. They do not include materials for you to use directly with your students, nor do they discuss classroom management, teaching styles, or assessment techniques. Instead, these books focus squarely on issues of mathematical content—the ideas and understanding that you must bring to your preparation, in-class instruction, one-on-one interactions with students, and assessment.

How do the authors approach the topics?

For each topic, the authors identify "big ideas" and "essential understandings." The big ideas are mathematical statements of overarching concepts that are central to a mathematical topic and link numerous smaller mathematical ideas into coherent wholes. The books call the smaller, more concrete ideas that are associated with each big idea *essential understandings*. They capture aspects of the corresponding big idea and provide evidence of its richness.

The big ideas have tremendous value in mathematics. You can gain an appreciation of the power and worth of these densely packed statements through persistent work with the interrelated essential understandings. Grasping these multiple smaller concepts and through them gaining access to the big ideas can greatly increase your intellectual assets and classroom possibilities.

In your work with mathematical ideas in your role as a teacher, you have probably observed that the essential understandings are often at the heart of the understanding that you need for presenting one of these challenging topics to students. Knowing these ideas very well is critical because they are the mathematical pieces that connect to form each big idea.

How are the books organized?

Every book in the Essential Understanding Series has the same structure:

- The introduction gives an overview, explaining the reasons for the selection of the particular topic and highlighting some of the differences between what teachers and students need to know about it.

- Chapter 1 is the heart of the book, identifying and examining the big ideas and related essential understandings.

Big ideas and essential understandings are identified by icons in the books.

marks a big idea, and

marks an essential understanding.

- Chapter 2 reconsiders the ideas discussed in chapter 1 in light of their connections with mathematical ideas within the grade band and with other mathematics that the students have encountered earlier or will encounter later in their study of mathematics.

- Chapter 3 wraps up the discussion by considering the challenges that students often face in grasping the necessary concepts related to the topic under discussion. It analyzes the development of their thinking and offers guidance for presenting ideas to them and assessing their understanding.

The discussion of big ideas and essential understandings in chapter 1 is interspersed with questions labeled "Reflect." It is important to pause in your reading to think about these on your own or discuss them with your colleagues. By engaging with the material in this way, you can make the experience of reading the book participatory, interactive, and dynamic.

Reflect questions can also serve as topics of conversation among local groups of teachers or teachers connected electronically in school districts or even between states. Thus, the Reflect items can extend the possibilities for using the books as tools for formal or informal experiences for in-service and preservice teachers, individually or in groups, in or beyond college or university classes.

A new perspective

The Essential Understanding Series thus is intended to support you in gaining a deep and broad understanding of mathematics that can benefit your students in many ways. Considering connections between the mathematics under discussion and other mathematics that students encounter earlier and later in the curriculum gives the books unusual depth as well as insight into vertical articulation in school mathematics.

The series appears against the backdrop of *Principles and Standards for School Mathematics* (NCTM 2000), *Curriculum Focal Points for Prekindergarten through Grade 8 Mathematics: A Quest for Coherence* (NCTM 2006), *Focus in High School Mathematics: Reasoning and Sense Making* (NCTM 2009), and the Navigations Series (NCTM 2001–2009). The new books play an important role, supporting the work of these publications by offering content-based professional development.

The other publications, in turn, can flesh out and enrich the new books. After reading this book, for example, you might select hands-on, Standards-based activities from the Navigations books for your students to use to gain insights into the topics that the Essential Understanding books discuss. If you are teaching students in prekindergarten through grade 8, you might apply your deeper understanding as you present material related to the three focal points that *Curriculum Focal Points* identifies for instruction at your students' level. Or if you are teaching students in grades 9–12, you might use your understanding to enrich the ways in which you can engage students in mathematical reasoning and sense making as presented in *Focus in High School Mathematics*.

An enriched understanding can give you a fresh perspective and infuse new energy into your teaching. We hope that the understanding that you acquire from reading the book will support your efforts as you help your students grasp the ideas that will ensure their mathematical success.

We wish to thank the reviewers of this volume who provided valuable feedback about the format and content of the book. In particular, Jane F. Schielack and David W. Henderson provided insights and suggestions that helped us present challenging ideas in accessible ways.

Introduction

This book focuses on ideas about geometry and measurement. These are ideas that you need to understand thoroughly and be able to use flexibly to be highly effective in your teaching of mathematics in prekindergarten–grade 2. The book discusses many mathematical ideas that are common in prekindergarten–early elementary grades curricula, and it assumes that you have had a variety of mathematics experiences that have motivated you to delve into—and move beyond—the mathematics that you expect your students to learn.

The book is designed to engage you with these ideas, helping you to develop an understanding that will guide you in planning and implementing lessons and assessing your students' learning in ways that reflect the full complexity of geometry and measurement. A deep, rich understanding of the breadth of the related concepts will enable you to communicate their influence and scope to your students, showing them how this kind of thinking permeates the mathematics that they have encountered—and will continue to encounter—throughout their school mathematics experiences.

The understanding of geometry and measurement that you gain from this focused study thus supports the vision of *Principles and Standards for School Mathematics* (NCTM 2000): "Imagine a classroom, a school, or a school district where all students have access to high-quality, engaging mathematics instruction" (p. 3). This vision depends on classroom teachers who "are continually growing as professionals" (p. 3) and routinely engage their students in meaningful experiences that help them learn mathematics with understanding.

Why Geometry and Measurement?

Like the topics of all the volumes in NCTM's Essential Understanding Series, geometry and measurement form a major area of school mathematics that is crucial for students to learn but challenging for teachers to teach. Students in prekindergarten–grade 2 need opportunities to think geometrically and reason spatially if they are to succeed in their subsequent mathematics experiences. Learners often struggle with aspects of geometric and spatial reasoning, not because they are incapable of them, but because they need frequent experiences and time to develop important thinking and visualization skills. For example, many students recognize a rectangle only when it has a standard position and shape (two long horizontal sides and two short vertical sides) and do not see how rectangles, squares, and rhombi are related. Teachers of mathematics in prekindergarten–grade 2 understand these particular aspects of geometry and are able to use their understanding to help students value classifications of geometric figures and to reason about classification systems.

Your work as a teacher of mathematics in these grades calls for a solid understanding of the mathematics that you—and your school, your district, and your state curriculum—expect your students to learn about geometric and spatial reasoning. Your work also requires you to know how geometry and measurement relate to other mathematical ideas that your students will encounter in the lesson at hand, the current school year, and beyond. Rich mathematical understanding guides teachers' decisions in much of their work, such as choosing tasks for a lesson, posing questions, selecting materials, ordering topics and ideas over time, assessing the quality of students' work, and devising ways to challenge and support their thinking.

Understanding Geometry and Measurement

Teachers teach mathematics because they want others to understand it in ways that will contribute to success and satisfaction in school, work, and life. Helping your students develop a robust and lasting understanding of geometry and measurement requires that you understand this area deeply. But what does this mean?

It is easy to think that understanding an area of mathematics, such as geometry and measurement, means knowing certain facts, being able to solve particular types of problems, and mastering relevant vocabulary. For example, you know such facts as, "A quadrilateral is a four-sided polygon." You are expected to be skillful in solving problems that involve length, area, and volume. Your mathematical vocabulary is assumed to include such terms as *classification, rhombus, unit,* and *congruence.*

Obviously, facts, vocabulary, and techniques for solving certain types of problems are not all that you are expected to know about geometry and measurement. For example, in your ongoing work with students, you have undoubtedly discovered that you need not only to know the names of common shapes but also to be able to understand and evaluate descriptions of these shapes that your students create.

It is also easy to focus on a very long list of ideas about geometry and measurement that all teachers of mathematics in prekindergarten–grade 2 are expected to know and teach. Curriculum developers often devise and publish such lists. However important the individual items might be, these lists cannot capture the essence of a rich understanding of the topic. Understanding geometry and measurement deeply requires you not only to know important mathematical ideas but also to recognize how these ideas relate to one another. Your understanding continues to grow with experience and as a result of opportunities to embrace new ideas and find new connections among familiar ones.

Furthermore, your understanding of geometry and measurement should transcend the content intended for your students. Some of the differences between what you need to know and what you expect students to learn are easy to point out. For instance, your understanding of the topic should include a grasp of formal properties of isometries—mathematics that students will encounter later but do not yet understand.

Other differences between the understanding that you need to have and the understanding that you expect your students to acquire are less obvious, but your experiences in the classroom have undoubtedly made you aware of these differences at some level. For example, how many times have you been grateful to have an understanding of geometry and measurement that enables you to recognize the merit in a student's unanticipated generalization or claim, such as, "A square is a rhombus!" How many other times have you wondered whether you could be missing such an opportunity or failing to use it to full advantage because of a gap in your knowledge?

As you have almost certainly discovered, knowing and being able to do familiar mathematics are not enough when you're in the classroom. You also need to be able to identify and justify or refute novel claims. These claims and justifications might draw on ideas or techniques that are beyond the mathematical experiences of your students and current curricular expectations for them. For example, you may need to be able to refute erroneous claims, such as, "A diamond isn't a square because it stands on a point." Or you may need to explain to a student who is using a ruler to measure the length of an object that the number at one end of the object represents the length if the other end of the object is at 0 on the ruler. Otherwise, the student will have to count the number of units between the number at the beginning of the length and the number at the end of the length—or find the difference between the two by subtracting.

Big Ideas and Essential Understandings

Thinking about the many particular ideas that are part of a rich understanding of geometry and measurement can be an overwhelming task. Articulating all of those mathematical ideas and their connections would require many books. To choose which ideas to include in this book, the authors considered a critical question: What is essential for teachers of mathematics in prekindergarten–grade 2 to know about geometry and measurement to be effective in the classroom? To answer this question, the authors drew on a variety of resources, including research on mathematics learning and teaching, the expertise of colleagues in mathematics and mathematics education, the reactions of reviewers and professional development providers, ideas from curricular materials, and personal experiences.

As a result, the mathematical content of this book focuses on essential ideas for teachers about geometry and measurement. In particular, chapter 1 is organized around four big ideas related to this important area of mathematics. Each big idea is supported by smaller, more specific mathematical ideas, which the book calls *essential understandings.*

Benefits for Teaching, Learning, and Assessing

Understanding geometry and measurement can help you implement the Teaching Principle enunciated in *Principles and Standards for School Mathematics.* This Principle sets a high standard for instruction: "Effective mathematics teaching requires understanding what students know and need to learn and then challenging and supporting them to learn it well" (NCTM 2000, p. 16). As in teaching about other critical topics in mathematics, teaching about geometry and measurement requires knowledge that goes "beyond what most teachers experience in standard preservice mathematics courses" (p. 17).

Chapter 1 comes into play at this point, offering an overview of geometry and measurement that is intended to be more focused and comprehensive than many discussions of the topic that you are likely to have encountered. This chapter enumerates, expands on, and gives examples of the big ideas and essential understandings related to geometry and measurement, with the goal of supplementing or reinforcing your understanding. Thus, chapter 1 aims to prepare you to implement the Teaching Principle fully as you provide the support and challenge that your students need to develop robust understanding of geometry and measurement.

Consolidating your understanding in this way also prepares you to implement the Learning Principle outlined in *Principles and Standards:* "Students must learn mathematics with understanding, actively building new knowledge from experience and prior knowledge" (NCTM 2000, p. 20). To support your efforts to help your students learn in this way, chapter 2 builds on the understanding of geometry and measurement that chapter 1 communicates by pointing out specific ways in which the big ideas and essential understandings connect with mathematics that students typically encounter earlier or later in school. This chapter supports the Learning Principle by emphasizing longitudinal connections in students' learning about geometry and measurement. For example, as their mathematical experiences expand, students gradually develop an understanding of connections between measurement and rational numbers and become skillful in reasoning with generalizations about the role of unit in both instances.

The understanding that chapters 1 and 2 convey can strengthen another critical area of teaching. Chapter 3 addresses this area, building on the first two chapters to show how an understanding of geometry and measurement can help you select and develop appropriate tasks, techniques, and tools for assessing your students' understanding of shapes and their classes, movement and location in space, transformations of shapes, and measurement. An ownership of the big ideas and essential understandings related to geometric and spatial reasoning, reinforced by an understanding of students' past and future experiences with the ideas, can help you ensure that your classroom practice reflects the NCTM Process Standards (NCTM 2000)—as well as the Standards for Mathematical Practice outlined in the Common Core State Standards (Common Core State Standards Initiative 2010)—and supports the learning of significant mathematics.

Such assessment satisfies the first requirement of the Assessment Principle set out in *Principles and Standards:* "Assessment should support the learning of important mathematics and furnish useful information to both teachers and students" (NCTM 2000, p. 22). An understanding of geometry and measurement can also help you satisfy the second requirement of the Assessment Principle, by enabling you to develop assessment tasks that give you specific information about what your students are thinking and what they understand. For example, a task that asks students to measure a length with a unit and then predict the effect on the count (the measurement) when using a smaller (or larger) unit can reveal important information about children's understanding of the concept of unit and the unit's relationship to the measurement.

Ready to Begin

This introduction has painted the background, preparing you for the big ideas and associated essential understandings related to geometry and measurement that you will encounter and explore in chapter 1. Reading the chapters in the order in which they appear can be a very useful way to approach the book. Read chapter 1 in more than one sitting, allowing time for reflection. Absorb the ideas—both big ideas and essential understandings—related to geometry and measurement. Appreciate the connections among these ideas. Carry your newfound or reinforced understanding to chapter 2, which guides you in seeing how the ideas in chapter 1 are connected to the mathematics that your students have encountered earlier or will encounter later in school. Then read about teaching, learning, and assessment issues in chapter 3.

Alternatively, you may want to take a look at chapter 3 before engaging with the mathematical ideas in chapters 1 and 2. Having the challenges of teaching, learning, and assessment issues clearly in mind, along with possible approaches to them, can give you a different perspective on the material in the earlier chapters.

No matter how you read the book, let it serve as a tool to expand your understanding, application, and enjoyment of geometry and measurement.

Geometry and Measurement: The Big Ideas and Essential Understandings

To recognize objects that we see and move safely and competently through the space that we live in—maintaining balance and recalling where things are in relation to one another—we're wired, from birth, for making sense of the space around us, and for *exploring* that space curiously to learn more about it. Even in infancy, we humans are remarkably sophisticated processors of shape and size! Babies who have learned to recognize a bottle when it is handed to them in the expected orientation do not at first reach for it when it is presented in an unfamiliar orientation. But in just weeks they learn to recognize objects regardless of their orientation.

This seems so natural to us—and it *is* so natural, a product of our evolution—that we hardly notice what a remarkable feat it is. The infant must not only be able to rotate the image of the bottle mentally but even account for changed size relationships within it. Consider the world from the infant's perspective. Viewed one way, the nipple takes up nearly half the length of the bottle, tip to base:

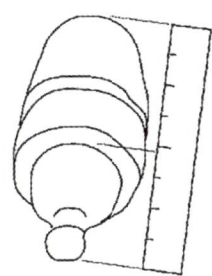

Viewed another way, it makes barely more than an eighth of the length:

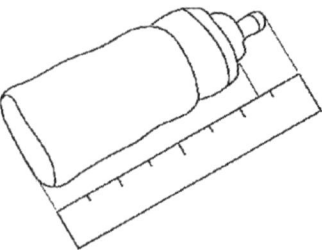

Infants also recognize certain symmetries. One-year-olds experiment endlessly with the inside-outside relation by putting objects into bags or boxes and taking them out again. Their equally tireless enjoyment of the game of peekaboo reflects the fact that they are fascinated by things disappearing and reappearing from behind or under other things.

Long before preschool, children develop an impressive system for recognizing structure in the space in which they live. They classify objects by abstract shape and size without any instruction from us, having no difficulty recognizing a tiny Lego figure as a person, despite its miniscule size and the many ways—including the fingerless shape of its "hands"—in which it is merely an abstraction of human form. They abstract form in other ways as well, being much more inclined to hold a block this way,

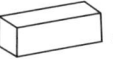,

rather than this way,

,

when they are pushing it around on the floor as a make-believe car or train. When they are pretending that it is a tower, they reverse that orientation, despite the fact that the block doesn't look the least bit like either a car or a tower. They are seeing abstract structure—a vastly simplified shape and an orientation.

Some modern geometries deal with such complex shapes as clouds, ferns, and fjords—these are "geometric" shapes, too. But recognizing the much simpler shapes of cylinders, spheres, and cones helps artists model the essential elements of the human form, which they often do quite abstractly before refining it to make it more lifelike than Lego-like. Clearly, children who recognize the Lego figure also see these simple geometric shapes as modeling the essential structure of a person.

Geometry—all of mathematics, really—is just an extension of what we do naturally, a systematization and refinement of our natural inclination to find structure. As geometers, we look for ways to classify and measure shape, orientation, and location. We look for attributes of shapes and positions, and we look for relationships among them. Then we move to a different level of sophistication by applying the same thinking to the attributes of shapes that we originally applied to the shapes themselves. For example, once we have noticed a special feature (like parallel sides) in a particularly familiar set of shapes (like rectangles), we may then think about what *other* shapes have parallel sides. We may wonder what that implies about their other features (like their angles), what happens when parallel lines are "cut" by other lines, whether *all* surfaces have parallel lines or whether they exist only on flat surfaces like the surface of a piece of paper, and so on.

The teacher's job is to understand enough about how geometric ways of thinking can extend and systematize the remarkable strengths that nature has given all students "for free." Teachers who have this understanding can feed the natural curiosity of their students and help them refine the analysis of their world that they are already actively, eagerly, and naturally engaged in.

Identifying the Big Ideas and Essential Understandings

In the remainder of this chapter, we focus on four interrelated big ideas and eleven associated essential understandings that support the big ideas. The big ideas highlight young children's perceptions of the world through a geometric lens. The essential understandings develop and refine each of the big ideas, unpacking the meanings, forms, and purpose of geometry that are important for teaching in prekindergarten–grade 2. These are ideas that teachers need to understand and be able to use flexibly to support their students' development of geometric concepts.

The big ideas and essential understandings are identified as a group below to give you a quick overview and for your convenience in referring back to them. Read through them now, but do not think that you must absorb them fully at this point. The chapter will discuss each one in detail.

Big Idea 1. A classification scheme specifies for a space or the objects within it the properties that are relevant to particular goals and intentions.

Essential Understanding 1*a*. Mathematical classification extends and refines everyday categorization by making more precise what we mean by "sides," "angles," "straightness," or other features that we attend to as we categorize mathematical objects.

Essential Understanding 1*b*. We may classify the same collection of objects in different ways.

Big Idea 2. Geometry allows us to structure spaces and specify locations within them.

Essential Understanding 2*a*. To describe a location, we must provide a reference point (an origin) and independent pieces of information (often called *coordinates*) indicating distance and direction from that point.

Essential Understanding 2*b*. Geometry and measurement can specify directions, routes, and locations in the world—for example, navigation paths and spatial relations—with precision. Given a reference point and an orientation, we can label position with numbers.

Essential Understanding 2*c*. Geometric objects are things that exist in our minds. Many of them are idealizations of things that also exist in the physical world.

Big Idea 3. We gain insight and understanding of spaces and the objects within them by noting what does and does not change as we transform these spaces and objects in various ways.

Essential Understanding 3*a*. Transformations can be used to describe differences between an idealized image of an object and the way that it is positioned in space or seen by the eye.

Essential Understanding 3*b*. Under each transformation, certain properties are invariant.

Big Idea 4. One way to analyze and describe geometric objects, relationships among them, or the spaces that they occupy is to quantify—measure or count—one or more of their attributes.

Essential Understanding 4a. Measurement can specify "how much" by assigning a number to such attributes as length, area, volume, and angle.

Essential Understanding 4b. Some quantities can be compared or measured directly, others can be measured indirectly, and the measurements of some objects are computed from other measurements.

Essential Understanding 4c. Measurement can be performed with a variety of units. The size of the unit and the number of units in the measure are inversely related to each other.

Essential Understanding 4d. Objects can be decomposed and composed to facilitate their measurement.

Classifying Objects: Big Idea 1

Big Idea 1. *A classification scheme specifies for a space or the objects within it the properties that are relevant to particular goals and intentions.*

Sorting shapes and giving names to the resulting collections are interrelated activities that are often treated as routine and simple—a bit arbitrary, and perhaps just a matter of memorizing fancy words. But they are not arbitrary at all. We don't invent ways of sorting things just because we can. We first discover that certain ways of grouping objects are useful for some purpose, and afterward we give that class of objects its own name and try to be precise about what criteria to use for that classification.

A discussion of names of geometric figures appears in *Developing Essential Understanding of Geometry for Teaching Mathematics in Grades 6–8* (Sinclair, Pimm, and Skelin 2011).

Once classification leads us to identify a collection of figures with some common features and calls special attention to those features, we need language to help convey this new knowledge. Clarity and precision of communication—not fussiness about terminology—is the goal of mathematical vocabulary. We use certain words not because mathematicians say we should but because they help us to be clear about what we're saying. This is the reason, after all, that they were invented.

For example, though angles come in all sizes, we give no special name to most of them. An angle of 77 degrees has no special name of its own but is lumped in with all other nameless angles smaller than 90 degrees; an angle with $136\frac{1}{2}$ degrees is similarly just lumped in a large category. But an angle of 90 degrees appears in so many important contexts that we need to be able to talk about it, so it does merit its own special name: *right angle.* And, because that particular angle is so important, we also need to talk about, and thus name, angles that are smaller than 90 degrees (*acute angles*) and larger than 90 degrees (*obtuse angles*). Reflect 1.1 invites you to think about how we communicate, and might communicate more clearly and precisely, about "side."

Reflect 1.1

What is a "side" of a two-dimensional shape, such as a rectangle?

How does your idea of a "side" of a rectangle fit with a sector based on a quarter circle?

If the "sides" of a rectangle are the line segments that form the rectangle's "edges," how are we to think about "sides" of a cube?

What might we mean if we talk about a cubic room's "sides" or "corners"?

The act of sorting focuses attention on attributes and therefore refines and extends what we know about objects. Classifying—sorting into classes and naming those classes—is an act of reasoning and logic that helps us distinguish objects that appear, superficially, to be the same. At the same time, it also helps us to see commonalities among objects that do not on the surface appear related. Having names for broad classes, and names for special sub-divisions within those broad classes, helps us communicate. Sorting and naming are useful!

Specifying objects precisely

Essential Understanding 1a. *Mathematical classification extends and refines everyday categorization by making more precise what we mean by "sides," "angles," "straightness," or other features that we attend to as we categorize mathematical objects.*

Classification and definition are closely related. A precise specification of the way that we are classifying a collection of objects gives us a definition for the objects in that collection. Initially, we start with experience—we get a rough sense of what we're talking about from examples—and then we try to make it precise. At first, we think of a triangle as a plane (flat) figure with three sides and three angles. At some point, we realize that's not precise enough, because that classification includes things like

,

for example, so we tighten the definition. Reflect 1.2 gives you an opportunity to consider how you might think about what a triangle is in responding to a common textbook task, as shown in figure 1.1.

Reflect 1.2

An exercise like that shown in figure 1.1 is commonplace in primary materials. The instructions might read, "Draw a ring around the pictures that have the shape of a triangle." The sample has a ring around the clothes hanger, and a child has drawn rings around six more pictures. Which of these pictures would you ring?

Fig. 1.1. A typical textbook task on triangles for young children

Understandably, most people would agree with the child, and what the child did is exactly what the curriculum writers had in mind. Artists use exactly this kind of "seeing" in quick sketching: arms are roughly cylindrical, heads are roughly elongated spheres, and sandwich halves are roughly triangular. But if the goal is to *build* rather than apply the concept of triangle, and to support recognition of the critical features of triangles, this exercise causes some mischief. Triangles are figures with three straight sides, but the sandwich halves have only one straight side. Triangles have three angles—that's what their name means—but the corners of the clothes hanger and the musical triangle are rounded, so they have no angles (and the musical triangle isn't closed). Further, all real objects are three-dimensional, so a two-dimensional figure is a creation of our minds—a mathematical idealization of the real world but not a *physical* thing that we can pick up or see. In fact, the toy block and the tent are drawn in ways that clearly emphasize their three-dimensionality, making them poor and misleading examples of a triangle. Calling them "triangles" is like calling a sphere or ball a "circle."

We don't, of course, nitpick with kindergarteners. If they say "circle" to describe the shape of a ball, we don't punish them or even interrupt to make them say "sphere." They are correctly recognizing an important feature of the figure and even using a correct name for that feature. Similarly, if they recognize triangular form in things that are not actually triangles, that's not only OK but excellent. But we—teachers and textbook writers—must not teach with examples or explanations that generate muddy or incorrect ideas. Mathematics is interested in content, such as shape and number, of course. Yet at its core, it is about precision of reasoning and thinking, which is not reflected in those "examples" of triangles. If we accept loose descriptions of triangles, we lose the certainty that mathematics offers.

For instance, if a "triangular" shape has two straight sides and a "curved part," the angle measures do not necessarily add to 180 degrees. That is why our definition refers to three sides, and the definition of *side* is a (straight) line segment. Once we are precise, we can be sure that anything we find out about triangles in a plane—such as the sum of the angles, or the fact that the sum of the lengths of any two sides is greater than the length of the third side—is true of every triangle in a plane. Similarly, once we define *square*, *rectangle*, and *rhombus* so that a square is both a rectangle and a rhombus, then we know that anything we prove about a rectangle (for example, that the diagonals are the same length) or a rhombus (for instance, that the diagonals are perpendicular to each other) must also be true about a square.

For precision, other words must be clarified as well. For example, a side of a polygon (a closed plane figure with all straight sides) does have to be straight. But what does this mean? People say that a "side" of a building is "straight," but this context uses "side" and "straight" with meanings quite different from the ones they had in the context of polygons. People also use "straight" to refer to honesty, clothes hanging close to the body, an unbroken period of time ("four straight wins" or "come straight home"), and more. Reflect 1.3 lets you explore the meaning that you give to "straight" in geometry.

Reflect 1.3

How can you describe exactly what is meant by "straight" in *geometry*? Find at least two ways to express this idea.

"Straight" might be defined as the shortest distance between two points or extending in the same direction without curving or bending. We usually think of this attribute on a plane, which is how we mean it here. But consider: If you could drive a car straight through a vast plain, thousands of miles, would you travel a straight line? If that vast plain covered the entire planet, you'd wind up back where you started if you kept driving straight long enough—and if the planet were perfectly spherical! Yet, we *do* refer to that path as "straight" on the earth's curved surface. On a plane, any path that returns to its starting place cannot be straight. This is one reason why we specified that the triangles that we are discussing are in a plane. Precision is necessary.

This discussion brings us back to the notion of idealization. Showing an equilateral triangle with a horizontal base, such as

can be a useful idealization. But without varied examples, we can't grasp the breadth of the idea. Once we have a precise definition of a triangle, we need to apply it to varied examples, such as

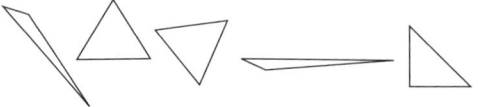

so that we see what are the defining attributes (a three-sided polygon) and what attributes are not defining (for example, bases do not have to be horizontal, sides do not have to be equal length).

To do this well, we also need *non-examples*. When a baby sees her first "doggie," she typically generalizes "doggie" to all small furry things, including cats. Perhaps, likewise, all pointy shapes are "triangles"! Seeing and naming other shapes might help a bit—after all,

are called "squares," so at least some pointy things are not called triangles. Although this is some help, it is not enough. Students will eventually learn that those squares can also be called "rectangles," so perhaps they can be called "triangles," too. Students need "near-miss examples"—examples that *almost* have the right features but not quite—to begin attending to the essential features. This process actually conflicts with the goal of idealizing. If we are to ignore the nicks and scrapes when we identify the door as rectangular, shouldn't we ignore the slight overlaps on the corners of a "triangle" like

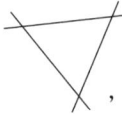

too?

It depends. If we are describing the shape for artistic purposes, or sketching it as part of a construction diagram—then, good enough, it's "a triangle." If we are trying to do mathematics—that is, analyze its properties so that we can come to conclusions that we can build on—then it isn't really a triangle, because its sides don't meet perfectly. We can idealize the pictures

as triangles with some kind of decoration or variation, but none are actual triangles (although parts of some of them are). Reflect 1.4 asks you to consider properties of examples and non-examples of triangles.

Reflect 1.4

Using just these pictures of triangles,

and these pictures of non-triangles,

and trying not to use any of the prior knowledge that you have as an educated adult, what can you *abstract* about what is or is not a "triangle"? That is, what is true of all five pictures called "triangles" that is not true of the eight pictures called "non-triangles," and how might you use this information to tell which of these pictures,

shows a triangle?

Classifying objects in multiple ways

Essential Understanding 1*b*. *We may classify the same collection of objects in different ways.*

If someone asked us to classify paths, we might classify them as "straight" or "not straight," depending on the context. Similarly, geometric shapes can be classified in many ways. Consider the collection of shapes in figure 1.2. Figure 1.3 shows the same shapes in and below a grid after a sorting of them. One way to classify the shapes is to sort them into categories that reflect common usage, as Reflect 1.5 explores.

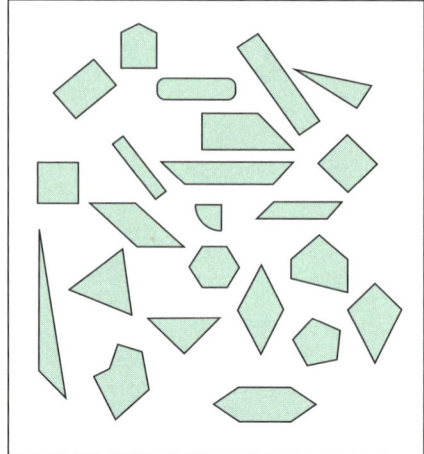

Fig. 1.2. A collection of geometric shapes

Reflect 1.5

Figure 1.3 shows the collection of shapes in figure 1.2 again, this time after sorting and classification in a grid. What would you call those shapes in the grid? Which of the shapes below the grid could you include in the categories already started?

How would you justify your way of classifying those shapes?

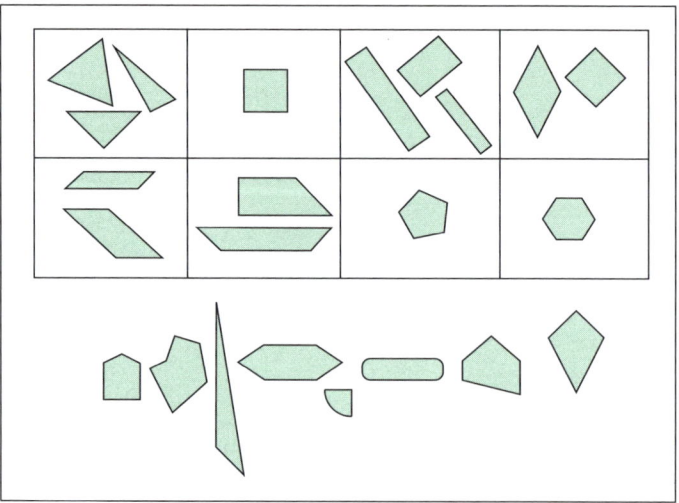

Fig. 1.3. The collection of shapes from figure 1.2, partially sorted

Another way to classify the shapes would be by choosing a single specific attribute. For example, the number of sides is the chosen attribute for the classification shown in figure 1.4, with the "leftovers" at the bottom. Reclassifying objects according to new criteria requires that we move beyond what we are used to and re-organize our reasoning at a higher level of abstraction. We could, if we had a suitable purpose, reclassify them by whether or not they have at least one right angle, as in figure 1.5.

Fig. 1.4. The shapes from figure 1.2 categorized by number of sides

Fig. 1.5. The shapes in figure 1.2 classified by "at least one right angle"

These are ways to examine such attributes and what they mean. A typical way that people describe rectangles to children is to say that they have "two long sides and two short sides." Figure 1.6 shows a classification of our shapes using those criteria. The shapes with two long sides and two short sides are at the top.

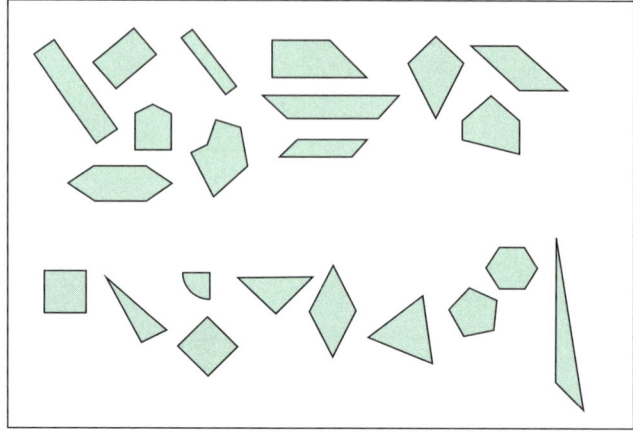

Fig. 1.6. The shapes in figure 1.2 classified by "two long sides and two short sides"

Hmm. This is not what we were after! Some of those shapes have two long sides and two short sides, but other parts as well. Perhaps we should have said "*exactly* two long sides and two short sides." Figure 1.7 shows the resulting classification.

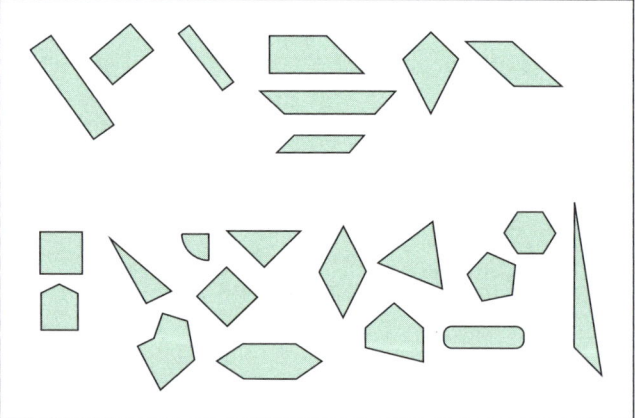

Fig. 1.7. The shapes in figure 1.2 classified by "exactly two long sides and two short sides"

This is better, but still not quite what we wanted. Of course! We should have said "exactly two long sides of the *same length* and two short sides of the *same length*." This change produces the results shown in figure 1.8. Reflect 1.6 offers an opportunity to investigate this classification further.

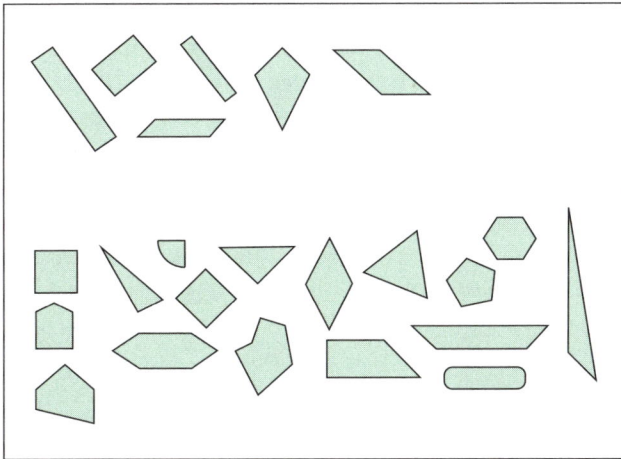

Fig. 1.8. The shapes in figure 1.2 classified by "exactly two long sides of the same length and two short sides of the same length"

Reflect 1.6

In the classification in figure 1.8, the shapes that meet our requirement (exactly two long sides of the same length and two short sides of the same length) include shapes that are not what we think of as *rectangles*. Are they all rectangles or not? What do we need to change in our criteria for rectangles?

Notice that "exactly two long sides of the same length and two short sides of the same length," a common description of rectangles, is still inadequate, and so it does not serve as a *definition* for rectangle. Once we add "all right angles" to our criteria, we might be satisfied. As figure 1.9 shows, all the objects in our set now are what we think of as *rectangles*.

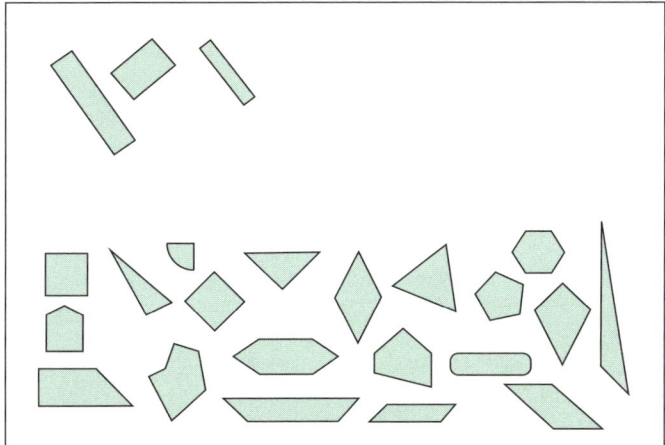

Fig. 1.9. The shapes in figure 1.2 reclassified after adding "all right angles" to the criteria

But look at the set of leftovers. Mathematicians might ask, Why omit the squares, *which have all the important properties of the rectangles?* We object that squares do not have the property we started with: two long sides and two short sides. Is that mathematically important? Why? Did it help us generate other useful properties of rectangles? No. We might want to reconsider our initial property about side lengths and choose to include squares as rectangles. Just as a German shepherd dog has all the characteristics of "dog," but also some additional characteristics, squares are a special breed, not a completely different species.

Such inclusive classification has power. Once we know that a basenji is a dog, we know a lot about it (for example, it is a domesticated carnivorous mammal, usually with a long snout, an acute sense of smell, and barking or howling vocalizations), even if we don't know its special characteristics (for instance, its vocalizing can sound like yodeling). Similarly, as we have seen, once we know that a square is a special type of rectangle, everything we know about rectangles applies to all squares. Classifications often involve hierarchies. This is the focus of Reflect 1.7.

Reflect 1.7

It is common for children—and adults, for that matter—to treat squares and rectangles as *separate* categories, but in fact all squares are rectangles. Not all rectangles are squares, of course, and this sets up a hierarchy: Rectangles are a broader category than squares; squares are a special kind of rectangle.

Find another example, not from geometry or even mathematics, in which *all A are B,* but *not all B are A.*

It is useful to classify geometric figures accurately and inclusively. Figure 1.10 shows a classification scheme for quadrilaterals based on the properties most commonly studied in elementary geometry. Figure 1.11 does the same for triangles. Reflect 1.8 invites you to interpret the classification scheme in figure 1.10.

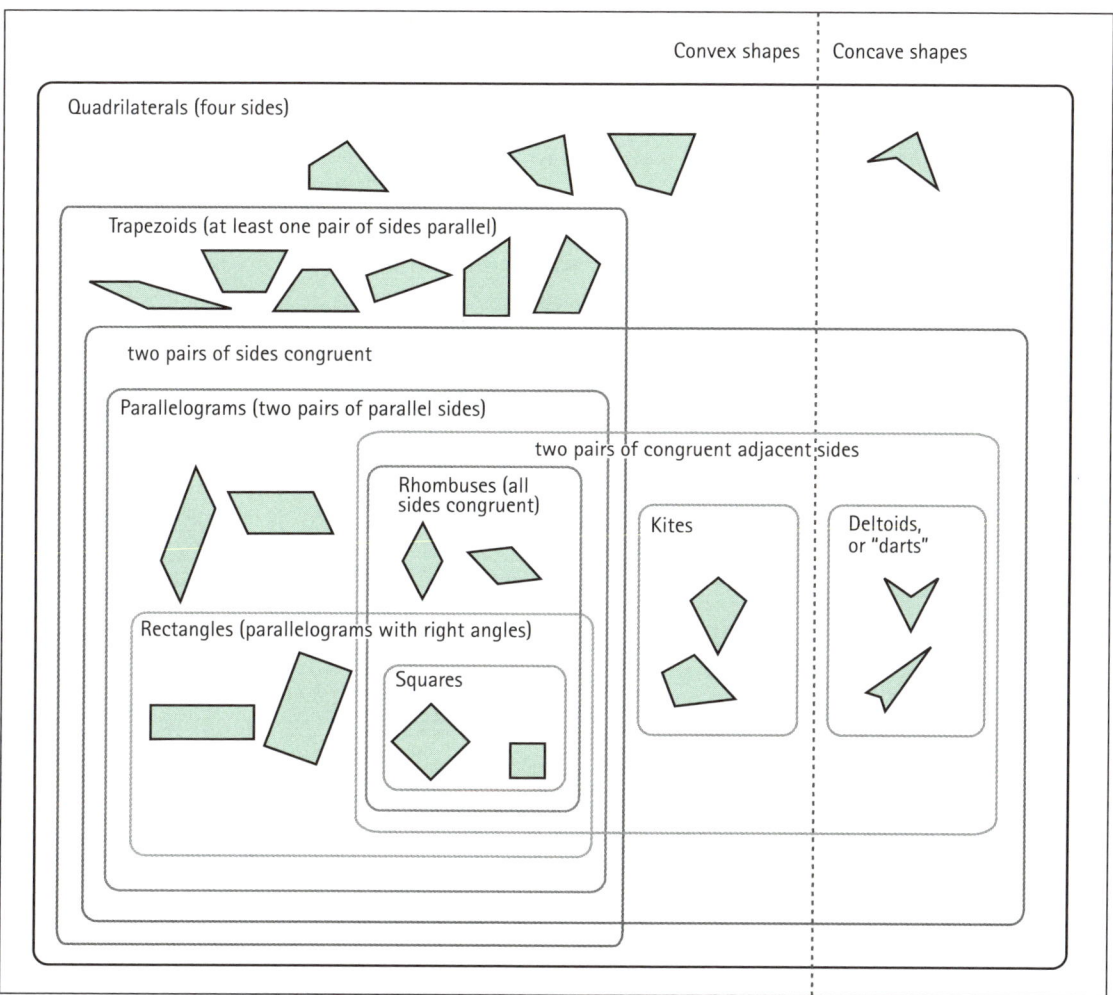

Fig. 1.10. A Venn diagram of quadrilaterals

Reflect 1.8

Figure 1.10 is a Venn diagram illustrating the classification of quadrilaterals.

How does it show that a square is a special type of rectangle?

For what class (or classes) is a rectangle a special type? What characteristics make it a part of that class? What characteristics make it a *special* part of that class?

In figure 1.10, notice that most of the categories are nested—not separate islands. In particular, trapezoids aren't classified as figures with *only* one pair of parallel sides, but as figures with *at least* one pair of parallel sides, so that parallelograms are "special trapezoids"—that is, shapes that meet all the criteria for trapezoids and some more. Squares are the most special of all. They have at least one pair of sides parallel—in fact, two—with all sides *congruent* (matching up perfectly), and right angles. Children don't need to learn all these names, but the idea of classification is important and, of course, extends well beyond mathematics.

Figure 1.11 shows a similar Venn diagram of triangles, and Reflect 1.9 asks you to consider the structure and importance of the names associated with types of triangles.

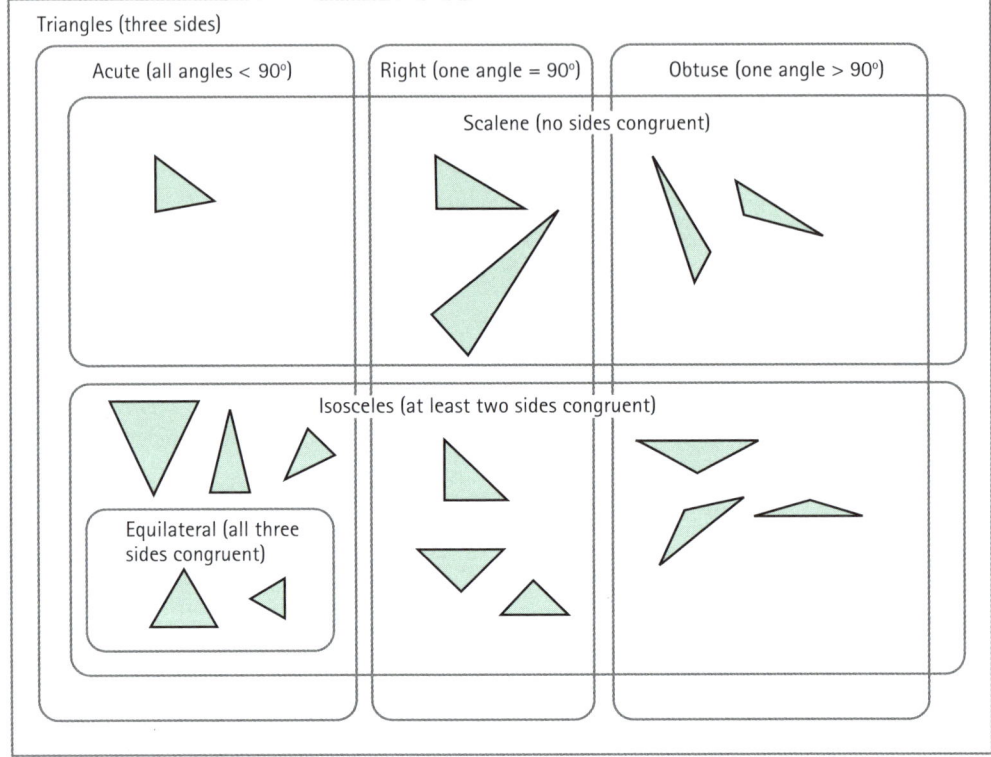

Fig. 1.11. A Venn diagram of triangles

Reflect 1.9

Figure 1.11 classifies triangles both by congruence of sides and by the size of the angles. Is the classification complete?

We call a triangle "acute" if *all* angles are acute—that is, measure less than 90 degrees. Why do we call a triangle "obtuse" if even just *one* angle is obtuse—that is, measures more than 90 degrees? Why would we not require all three, or at least a majority, of the angles to be obtuse?

The classifications in the two Venn diagrams have a lot of structure. For both the triangles and the quadrilaterals, certain properties are used to distinguish objects, usually into two classes: those *with* that property, and those *without* it. In the case of triangles, the property "having a right angle" divides triangles into two classes: right triangles and non–right triangles. But subdividing non–right triangles into acute triangles and obtuse triangles is useful. Classification helps us reason about objects by a sort of divide-and-conquer approach. We find that thinking about what's true of *all* quadrilaterals is too hard at first and also limits what we can say because quadrilaterals are so varied.

Classification also helps us communicate accurately about mathematical objects such as geometric figures. Mathematical classifications allow us to differentiate among objects that superficially appear similar and to see commonalities among objects that do not initially seem related. Classifications allow us to use reasoning such as, "Because the diagonals of all parallelograms bisect each other, the same must be true of the diagonals of all squares, because a square is a special type of parallelogram."

Structuring Space and Identifying Location: Big Idea 2

Big Idea 2. *Geometry allows us to structure spaces and specify locations within them.*

Children are sensitive to both location and distance at an early age, even if their knowledge about the formalities of measurement is very undeveloped. Spatial relationships—including relationships between shapes and sizes—matter. We measure objects to see which of two things is bigger when we can't tell by sight alone—when, for example, we want to know how much a plant has grown. We also measure distances to find out how far a trip might be, or to map out our neighborhood or the world, with coordinates, such as lines of longitude and latitude.

A preschooler whose height has been measured regularly at home might hold a ruler against another child, anywhere and in any direction, and simply declare "seven inches," or even just "seven." The child might also say "seven feets" or "seven pounds" for height but not "seven books," so the child has a *category* for measurement terms, even if measurement units are as yet un-learned. The child also knows to say a number but hasn't necessarily thought about what that number means or how the measurement tool gives that number.

Children's "measurement" of distances is similar. They have a notion that an uncle's house is farther away than their neighbor's house, with only a beginning notion of the measurement units, which could be "10 miles" or "10 more minutes." The measurement of *distances*, a process that helps us find where one object is in relation to another, begins our discussion of Big Idea 2.

Describing a location

Essential Understanding 2a. *To describe a location, we must provide a reference point (an origin) and independent pieces of information (often called coordinates) indicating distance and direction from that point.*

One key idea—the idea that lies behind teaching children to "line up the end of the ruler..."—is about more than measurement. To specify an object's location in space, we must start with a reference point, a known place (often called the *origin*), some information about direc-tion, and, if we need to be precise, a distance from the origin to the location. For instance, we might say, "The thing you're looking for is under your placemat" (or "three cars ahead of mine," or "twenty paces east of the oak tree and twelve feet under"). That is where the object is. Reflect 1.10 encourages you to examine the examples in the placemat, car, and oak tree settings more formally.

Reflect 1.10

Consider each of the following instructions:

- Look under your placemat.

- Look three cars ahead of mine.

- Look twenty paces east of the oak tree and twelve feet under.

What words specify the origin, direction(s), and distance(s) in each case?

Note that we could give just *one* distance from the base of the oak tree to the buried treasure if we specified the direction precisely enough. How might we do that?

To measure the length of a shoe, we match the shoe's "origin" (let's say the very back of the heel) and the ruler's "origin" (the end that would be marked "0" if there were room). We also align the ruler to match the direction from the origin (the back of the heel) to the other end of the shoe. The number that matches the toe tip's location is the length, in whatever units the ruler uses.

The number line is a ruler that runs in both directions from its origin, zero. The numbers indicate both direction and distance from the origin, with the convention that positive numbers indicate distances to the right. The number that is exactly one-half unit (a distance) to the right (a direction) of the origin is $1/2$. The number that is two units to the left of the origin is -2. That compact notation "-2" specifies both distance (2) and direction ($-$) from 0. We can even "measure" the distance between $1/2$ and 3. The distance between 0 and 3 is 3 units; the distance between 1 and 3 is 2 units, and the distance between $1/2$ and 3 is $2 \, 1/2$ units. Any number can be specified by its location in relation to another number, which suggests an interesting kind of puzzle that can be given to children to solve:

"I'm a number. I'm exactly 3 units away from 7. Who am I?"

"Hmmm... I'm not sure yet. Which *direction* from 7?"

"To the right."

"Oh! OK, I know!"

This number puzzle did not have a unique answer until we knew the direction as well as the distance and the starting point. We build on the relative locations of numbers on a number line or ruler to specify locations in the real world and to describe movement through the world around us.

Specifying real–world locations

Essential Understanding 2b. Geometry and measurement can precisely specify directions, routes, and locations in the world—for example, navigation paths and spatial relations. Given a reference point and an orientation, we can label position with numbers.

Geometric objects, along with their measurements, provide the structures for analyzing locations, directions, and coordinates. To specify the location of a point along a line, we need only a starting place (the origin), a direction from that origin along the line, and one measurement. We often use this idea in talking about real-world settings. For example, we sit two seats to the left of a friend at a long cafeteria table. The friend's seat is the reference point, "left" is the direction, and "two seats" is the one measurement.

A reference point and one measurement work well when we describe things that fall along a line. However, to describe a location in many venues requires more. For example, we can describe a particular seat in an auditorium starting from the front left corner of the seating area and noting that it is four rows back and seven seats in. The corner is our reference point, but we now need two measurements, each with a direction, to specify the location. Of course, we had some choice, such as choosing the left back corner as our starting point. Mathematically speaking, locating the seat on the side of the cafeteria table is like locating a number on a number line; locating the seat in the auditorium is like locating a point on a plane.

Work with a map also typically involves locating a point on a plane. Any work with maps involves four questions:

- Which way? (direction)
- How far? (distance)
- Where? (location)
- What objects? (identification)

Answering these questions may require processes of abstracting, generalizing, and symbolizing. Some map symbols are icons, such as an airplane for an airport, but others are more abstract, such as circles for cities, and some maps use grids composed of arrays of squares to help users find, for instance, a park located in "region D4." To find the park, users find the column labeled "D" and the row labeled "4." Their intersection defines a square region that contains the park.

Finding a park located in region D4 on a map uses the structure of rectangular arrays as a key way of analyzing two-dimensional space. Developing understanding of rectangular arrays is discussed in connection with Essential Understanding 3b.

To specify the location of a point on a plane, we still need only one starting place (the origin), but now we have choices. We can specify one distance from that origin if we are able to specify the direction precisely (say, as the compass direction 31 degrees north of east). But, as in the case of the park located in region D4 on the map, we more commonly use two perpendicular number lines, crossing at the origin (their zeros), to specify two directions—east-west and north-south—and two distances.

If we are locating a fly's position in a room, however, we may need three dimensions—three number lines—to specify, for example, how far the fly is from the blackboard wall, the window wall, and the floor.

Thus, we can use numbers to convey locations in different dimensions. We need one number for one dimension, determining, for example, the position of a point along a line (like a number line); two numbers to specify the location of a point in two dimensions; and three numbers for three dimensions. The number of dimensions that an object has is *defined* as the minimum number of coordinates (numbers, measurements) required to specify the location of any point in that object.

Very elementary experience with first one- and then two-dimensional space foreshadows not only the use of coordinates in three-dimensional space but also the use of other kinds of coordinates. For example, longitude and latitude provide a coordinate system for locations on the surface of the earth. Each latitude-longitude pair represents an intersection point of two number "lines" (circles on the surface of the earth). Because the two number lines can be made closer and closer together by choosing increasingly smaller units, the latitude-longitude pair can be as precise as necessary. For example, the location of Mount Everest might be approximated by using degrees as units (28°N 87°E) but can be made more precise by using minutes—sixtieths of a degree (27°59′N 86°57′E)—or even more precise by using seconds—sixtieths of a minute (27°59′16″N 86°56′40″E). Locating a corner of a house, which is much, much smaller than Mount Everest, requires more precision than degrees-minutes-seconds; the location given might include tenths or hundredths of a second.

Longitude and latitude can sometimes serve as a comfortable way to introduce the idea of coordinates to adults, because this system is a real-world example of numbers being used to specify locations. But it's not a familiar jumping-off point for children. Even though it uses the same basic idea as the familiar square-grid coordinate system that we use in much of mathematics, it is in many ways much more complicated. For one thing, the surface of the earth is (roughly) spherical, already much more complicated than the plane. As a result, all of the coordinates are specified as angle measure, not distance. Greater precision in those measures doesn't follow the familiar base-ten structure. Instead, it is a system of six-tieths that is a relic of the Babylonians.

The Cartesian coordinate system that we are most familiar with in school, and on which we plot points and draw graphs, does not work sensibly on a sphere but is perfect for the plane. It labels all position on the plane—every point—by referring to two perpendicular number lines—the axes—which intersect at their "0" positions, the origin. The use of a plane to describe a flat surface is an idealization of the flat surface, and one of many ways in which we use idealized abstract mathematical objects to describe objects in the real world.

Idealizing real-world objects

Essential Understanding 2c. *Geometric objects are things that exist in our minds. Many of them are idealizations of things that also exist in the physical world.*

We are used to illustrating geometrical ideas with physical objects. "The door is a rectangle," we say. What we mean, of course, is that the door's *shape* is a rectangle—ignoring thickness, and the doorknob, and the decorations on the door, and the chips along its edge, and the door's slight warping. What about a piece of paper? Well, it's rectangle shaped, but not "really" a rectangle. It, too, has thickness. And if we looked closely at its edges under a microscope, we would see that they're ragged, not straight lines at all. Even the drawings that we make on paper are not rectangles, triangles, and circles. The ink has thickness, and rectangles, triangles, and circles are just two-dimensional objects. We humans are trapped in a three-dimensional world. Anything that we can see is three-dimensional, even if one of those dimensions is so thin we can't perceive its thickness.

All the geometric shapes that we talk about are just idealizations—things that we can make in our minds but not in the physical world. But it's a good thing that we can make them in our minds! By simplifying a ragged edge to a straight one, and by ignoring chips, decorations, and doorknobs, we can analyze the abstract shapes and apply what we learn about them back to the physical world. Even better, we can apply what we learn to the next wonderful *non*-physical-world idea that we have, and invent something brand new. If we did not simplify and idealize, the complexity would be too burdensome, and we would not be able to solve very real problems. That is why things that exist only in our minds are such important parts of our very real world.

Our imaginations and dreams, although not part of our physical world, are very much a part of our real world. To build a house, we must picture in our minds before we can see with our eyes how the parts go together. We must make those mental pictures before we cut the wood, and maybe even keep notes of them on paper, or the parts simply won't fit.

Mathematics is part of the real world of our minds. It helps us sharpen and test the mental pictures, bringing them into focus and letting us juggle more and more complicated ideas.

Transforming Space and Objects: Big Idea 3

Big Idea 2. *We gain insight and understanding of spaces and the objects within them by noting what does and does not change as we transform these spaces and objects in various ways.*

Objects appear in various positions and orientations and at various distances. Despite all the differences that this makes in their appearance—for example, objects at a distance appear smaller than objects close up—our brains latch on to certain relationships among features that don't change.

Using transformations and images

Essential Understanding 3a. *Transformations can be used to describe differences between an idealized image of an object and the way that it is positioned in space or seen by the eye.*

We see pictures in many ways, sometimes directly in front of us, sometimes in mirrors, sometimes lying flat on the table, or sometimes distorted. Reflect 1.11 asks you to consider what happens when familiar letters appear in a mirror.

Reflect 1.11

Holding the letter **A** in front of a mirror can create the image shown here:

Suppose that you hold the letter **E** in front of a mirror in the same way. Describe the image that appears in the mirror.

Reflection transformations and symmetry

One common transformation that we see is our own image in a mirror. The image looks familiar, but if the viewer is holding up her right hand, her image is holding up its left hand!

How the mirror image does or does not differ from the original picture depends on properties of the original. For example, when we reflect the letter **A** in the mirror this way,

,

the image doesn't seem at all different from the original. Reflections like this exchange the left and right sides of the object. Shapes that have left-right symmetry, like the letter **A**, are unchanged in such a mirror, but shapes like the letter **E**, whose left and right halves aren't the same, will be changed:

By contrast, when we reflect the letters **A** and **E** in the mirror this way,

,

the **E** does not change, whereas the mirror image of **A** is upside down. Reflections like these exchange the top and bottom of a figure, so **E**, which is the same on top and bottom, appears unchanged. The properties of the original shape determine how it will respond to these two kinds of reflection.

Mirrors can be held at any angle and transform figures in interesting ways. Placing a mirror across a figure, like this,

,

covers up part of the figure and reflects only the other part, which can, of course, create a completely new shape. If the original figure is symmetric in certain ways, as the letter **A** is, then there is at least one way to make the mirror cut across it and still have a result that looks exactly like the original figure:

Figure 1.12 shows a design that has two dotted lines indicating *lines of symmetry*—lines that split the figure in half in a way that a

mirror, placed along one of them, would recreate the entire figure. Figure 1.13 shows that a mirror placed in some other direction on the figure also creates a symmetric picture—a mirror must do that— but one that is different from the original figure.

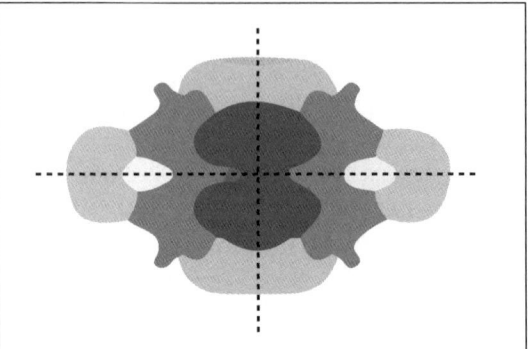

Fig. 1.12. A figure with two lines of symmetry

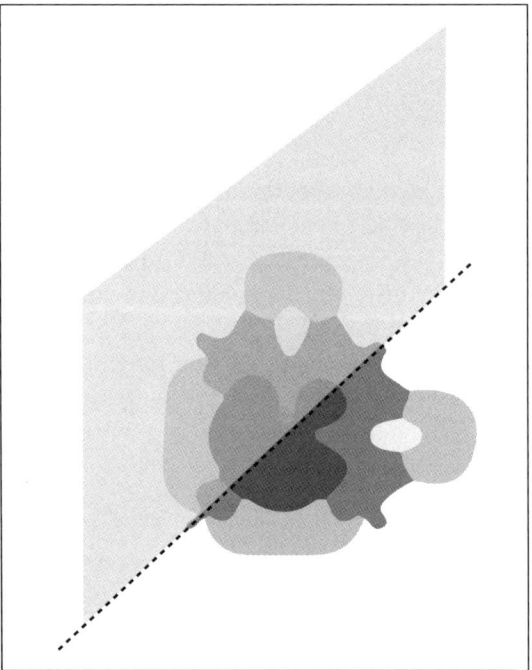

Fig. 1.13. The result of placing a mirror across a figure but not along a line of symmetry

It is particularly important to note that not every line that cuts a figure perfectly in half is a line of symmetry. The rectangle in figure 1.14 is cut exactly in half by the dotted line. The two sides are *congruent*—matching exactly, feature for feature.

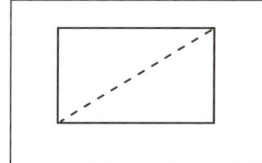

Fig. 1.14. A rectangle as two congruent triangles

Yet, the dotted line is not a line of symmetry. A mirror placed along that line does not reproduce the original figure. The mirror covers one triangular half of that rectangle and reflects the other triangular half so that the combined picture—what we see directly and what we see in the mirror—is symmetric. The result, shown in figure 1.15, is again made of two congruent triangles, the same two triangles that make up the rectangle, but this new figure is not the one we started with.

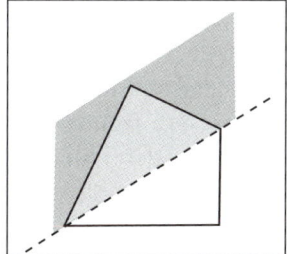

Fig. 1.15. A reflection that creates a new figure that is not the rectangle

In fact, the resulting figure is not a rectangle. It does have two long sides of the same length and two short sides of the same length. What does this tell us about that all-too-common description of rectangles? Reflect 1.12 provides an opportunity to explore the results of placing a mirror in different positions on a non-square rectangle.

Reflect 1.12

Find a small rectangular mirror. On a piece of paper, draw a non-square rectangle, smaller than the mirror.

How can you place your mirror on the rectangle so that the complete picture (in and out of the mirror) is still a rectangle but longer than the one you started with? Can you place the mirror in such a way that you make a rectangle that is skinnier than your original? Precisely square? A smaller square?

Which of the following non-rectangles can you make? A kite (see fig. 1.10)? A triangle? A five-sided figure? A six-sided figure? A seven-sided figure?

Is it possible to place the mirror so that you create two rectangles? Three rectangles?

When a mirror is placed in such a way that it does not cut a figure but reflects the entire figure, such as an entire rectangle, some things may change, but many things remain unchanged. For example, in

 ,

all the lengths and angles in the reflection are the same as the lengths and angles in the original. If the original is a triangle, or an **E**-shape, or a picture of a cat, so is the reflection, though it may face the other way. Reflections also don't change the area or perimeter of a figure. In fact, reflections don't change any measurement that we can make in the figure.

Transformations that change none of the measurements in a figure—that leave the measurements *invariant* (unchanged)—are called *isometries* (with *iso-* meaning "same," and *-metry* meaning "measure"). An isometry, therefore, creates an image that is congruent to—informally, the same size and same shape as—the original figure (called a *pre-image*). In fact, these transformations—the isometries—are conventionally used to define geometric congruence: two figures are defined as *congruent* if one or more isometries will transform one figure to coincide exactly with the other (*superposition*—lying one on top of the other to match precisely).

Rotation: Another isometry

Reflections are isometries. They preserve lengths and angles. In Reflect 1.13, determine what is changed and what is unchanged for a particular *rotation* of the shape shown in figure 1.16.

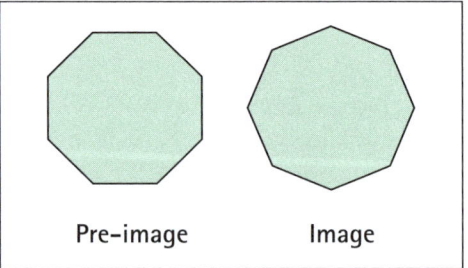

Fig. 1.16. An octagon rotated exactly $22^1/_2$ degrees (or $67^1/_2$ degrees, or $22^1/_2$ degrees + any multiple of 45 degrees)

Reflect 1.13

Look at figure 1.16. What stayed the same from the original (the *pre-image*) to the image? What changed?

Rotations, like reflections, preserve measurements: distance (and therefore angle) remains invariant when objects are rotated. Rotations just change the positions of objects in space. Informally, we think of a rotation as simply turning the object. For example, although people often perceive only the shape on the left in figure 1.16 as a "stop sign shape," further analysis shows that the two shapes have the same geometric properties and measures—they are congruent. Their position in space is different, but that is all.

However, we need to specify a center for the rotation, and an amount of rotation (in a specified direction). This is reminiscent of specifying an origin on a number line and a number that gives a distance (and direction).

At first, it can seem easy to confuse rotations and reflections. Consider, for example, the rotations of the letters **E**, **A**, and **F**, and a re-flected **F**, shown in figures 1.17, 1.18, 1.19, and 1.20, respectively.

Fig. 1.17. Rotations of **E**

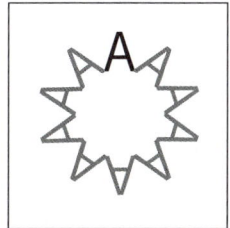

Fig. 1.18. Rotations of **A**

Fig. 1.19. Rotations of **F**

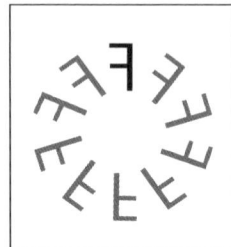

Fig. 1.20. Rotations of a reflected **F**

Figure 1.17 shows ten different rotations of the letter **E** around the center of a wheel. The top position is the original **E**. All the others are rotations of that **E**. The bottom one, however, looks exactly like the reflection of the **E** shown in the mirror in

.

If we perform the same experiment with the letter **A** (fig. 1.18), we note that the bottom image could also have been produced by a reflection, though a different one:

So is a reflection just one particular rotation? No, only when a figure has enough symmetry can its image under a reflection and a rotation turn out to be the same. The letter **F** has no line of symmetry, so no reflection of it will match a rotation of it. The top shapes in figures 1.19 and 1.20 are **F** and its reflection, and the remaining shapes in those figures are nine other rotations of that top shape. No result in figure 1.19 matches any result in figure 1.20. Rotations and reflections are different. We can rotate the **F** on the page to get any of the images shown in figure 1.19, but the only way to get from **F** to **ꟻ** with one transformation might be thought of as flipping the shape off the page into the third dimension and then setting it back down on the page. Rotations of that image, ⅂, will produce the shapes of figure 1.20.

In the preceding paragraphs, we consider the results of applying only one transformation. Reflect 1.14 offers an opportunity to extend this thinking to an important result that arises when a combination of two or more isometries of a particular type is applied.

Reflect 1.14

We are used to thinking of each type of transformation as having a completely distinct effect—informally, sliding a figure a particular distance (*translation*), flipping it over a particular line (*reflection*), or turning it by a particular angle around some point (*rotation*).

In fact, it is possible to get all three effects by using combinations of just *one* of these types of isometries, if we use that one in just the right way. That one type of move can be viewed as more "basic" than the others because a combination of two or more moves of that type can produce the same effect as one transformation by either of the other two isometries.

Which isometry do you think that is? How could it be used to produce the other two isometries?

Our visual system has evolved to allow us to recognize objects in whatever position they may be. Different shapes that hit our retina are perceived as the same object if the only difference among them is position. So, just as all the shapes in figure 1.19 appear to us as **F**s in different positions, all the flag-shaped outlines in figure 1.21 would naturally be perceived as "the same" object in different positions.

Fig. 1.21. Rotations of the pre-image and the reflection of a flag-shaped outline

When we learn to read, the rules change, and we must unlearn some of what we learned shortly after birth. Though the letter symbols for lowercase **p**, **b**, **d**, and **q** are just reflections and rotations of one another, we must nevertheless consider them as different when we read. Everything else continues to follow the old rule, with printed symbols as the exception. The letter reversals that young children often make are, therefore, not pathological at all. They show that the children's visual systems are operating exactly the way a competent visual system *should* operate if it is to save us from the tiger that might have turned to face a different way since we last looked. That is the way this system has evolved to operate since well before humans first appeared on the planet; reading, by contrast, has been around for no more than about five thousand years.

This hard-wired locking on an object's "sameness" while disregarding differences in its presentation has geometric consequences that are useful to know. Even our preschool children are experts at recognizing objects regardless of rotations or reflections. What they are not yet experts at is paying much attention to the fact that the rotation or reflection has happened!

Consider what typically happens when pre-readers play the puzzle game Meta-Forms (Lyons and Lyons 2007). Children as young as age 4 can approach this rich sequence of eighty puzzles, and adults will find the later puzzles in the sequence challenging. The player has nine shapes along with clues about where to place them on a 3 × 3 board. The shapes are triangles, squares, and circles, in three different colors—red, blue, and yellow in the actual puzzles. Unfortunately, those colors cannot be reproduced here. Figure 1.22 shows part of the clue page for the first puzzle. In the top row in the original, the triangle and square are red; in the bottom row, they are yellow.

Fig. 1.22. Meta-Forms board (Lyons and Lyons 2007)

In the page shown, the clue (top row, left)

suggests putting the red triangle (shown as gray here) in the middle of the left-hand column. The clue (top row, right)

suggests placing the red square (also shown as gray in the figure) in the bottom right corner. When all the pieces are placed and the board is filled, the puzzle is solved.

Some young children don't recognize the clues at all and simply place pieces where they like. Some see the entire clue page as a map of where the pieces go, and they put the red triangle in the top left-hand cell on the board because it is the top left picture on the clue page. And some, of course, place the pieces exactly where we adults would.

But some pre-readers who do recognize each clue as showing the spot where the piece is to be played may place the red square in any corner at all, effectively rotating the board. In other words,

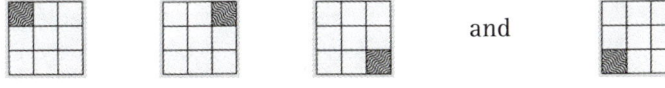

and

may all appear to them to be the same clue. The children may be inconsistent about the rotations, placing the yellow square (bottom row, left, in fig. 1.22, and shown as green) in some other corner, not necessarily opposite the red square. But, remarkably, some pre-readers are totally consistent, solving the puzzle entirely correctly but upside down, or perhaps switching left to right! After children learn to read, the rotations and reflections stop. And children who can't properly use the clue

because they still rotate never have trouble with

.

Transformations that do not preserve congruence

Reflection and rotation preserve distance and therefore congruence. As discussed earlier, any transformation that preserves congruence—thus, any transformation of the plane that leaves distance invariant (unchanging)—is called an isometry. Isometries preserve properties related to measurements, such as area (informally, the amount of "stuff" that covers a two-dimensional figure) and perimeter (informally, the distance an ant must crawl along the edge of the figure to get all the way around it) and, of course, angle measure.

Some actions or transformations on shapes do not preserve distances. These non–distance preserving actions do not always give congruent shapes. Among them are two important and commonly experienced transformations—*dilations* and *projections*—and ways of decomposing and recomposing shapes that are often called *dissections*. In school, projections are rarely discussed in mathematics class, but they are a major focus of attention (without the mathematical details) in perspective drawing.

Informally, we can think of dilations as enlargements or reductions of a shape. Think of a copy machine that can enlarge or shrink a figure. Slightly more formally, we can say that dilations "enlarge" the shape by a factor that can be greater than 1 (a true enlargement), equal to 1 (no change), or less than 1 (a reduction in size). When we become even just a bit more formal, the enlargement suddenly presents us with something of a surprise. By formal definition, a *dilation* is a transformation of all points away from one fixed point, which we call the *center of dilation*, by a fixed factor of their original distance from that point. If the center is called P and some point is initially 3 inches from P, then after a dilation by a factor of 7, the image of that point will be 21 inches from P. Figure 1.23 illustrates a dilation around P by a factor of 2, showing its effect on the original, smaller **F**.

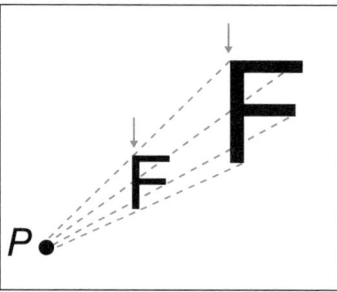

Fig. 1.23. Dilation of the plane by a factor of 2, with point P as the center of dilation, showing the smaller **F** transformed to the larger one

In figure 1.23, all the points on the plane are mapped to places twice as far from P as they were originally. In particular, all points on the bigger **F** are exactly 2 times the distance from P as corresponding points on the smaller **F**. Arrows mark the top left corner of the two **F**s to highlight the fact that one of them is twice as far from P as the other.

Although the definition of dilation specifies only the distances of points from the center of dilation, the result of the dilation in figure 1.23 is that the new **F**—its image—is also twice the size of the original **F**—the pre-image. Surprise! The points on the large **F** are not only twice as far from P as the corresponding points on the small **F** but also twice as far from each other as are the corresponding points. Although the actual distances change when we dilate, the proportions remain invariant. If we double some distances, we double them all.

The dotted lines in figure 1.23 also suggest the idea of projecting a light from P through the smaller **F**, and projecting that **F** on the wall—larger of course. Reflect 1.15 reverses the situation.

Reflect 1.15

What if the large **F** in figure 1.23 is the pre-image (the original figure) instead of the image?

How would you describe the dilation around *P* in this case?

Perhaps the best way to understand the transformations that alternate the small **F** and the large **F** as the pre-image is to think about zooming in or out on a scene with a camera, or doing the same on a map on your computer. The center of view, the center of dilation, remains fixed and objects shrink or enlarge as the some of the scene comes into or goes out of view.

A photograph captures the scene "exactly as you saw it," but doesn't preserve size. You can print it larger than life or smaller than life, changing the *size* of the objects but not what we informally call their "shape." The change in size that comes about as a result of a dilation (or photographic enlargement or reduction) preserves *angles* (that is, angle is invariant) but not *distances*. We call these figures that differ only in size *similar figures*.

In geometry, "similar" has a much more specialized meaning than it does in casual English. It does not capture just *any* property that is common among objects—such as, for example, the fact that all the objects in a set have exactly seven sides, or that they all are composed only of straight lines. Casual descriptions of congruence and similarity try to convey roughly the right idea without technical definitions. Informally, "congruent" often indicates "same shape and size," and "similar" suggests "same shape but possibly different size"—and these are adequate starting places. However, the problem with using these as *definitions* is that we have a very hard time nailing down precisely what we mean by "shape," and we don't have a much easier time stating precisely what we mean by "size."

When we need to be more precise, we just avoid those fuzzy terms altogether. In plane geometry, we say that two objects are *congruent* if one can be *transformed to the other*—that is, superimposed on the other—by any combination of isometries (reflections, rotations, translations), since these are moves that do not change distances. For example, in figure 1.24, C and D are *congruent* to A: C results from a reflection and a translation of A, and D results from rotating C clockwise 90 degrees and translating it (or reversing the order of these isometries). We also say that two objects are *similar* if a dilation (perhaps with a combination of isometries) of one is congruent to the other. In figure 1.24, B and E are *similar* to A: B is a dilation of A by a factor of $-(1/2)$, and E is a rotation of A by 180 degrees combined with a dilation by a factor of 2. Because a dilation can be by a factor of 1 (neither zooming in nor zooming out), objects that are congruent are also similar. In figure 1.24, all the shapes are similar, and A, C, and D are also congruent.

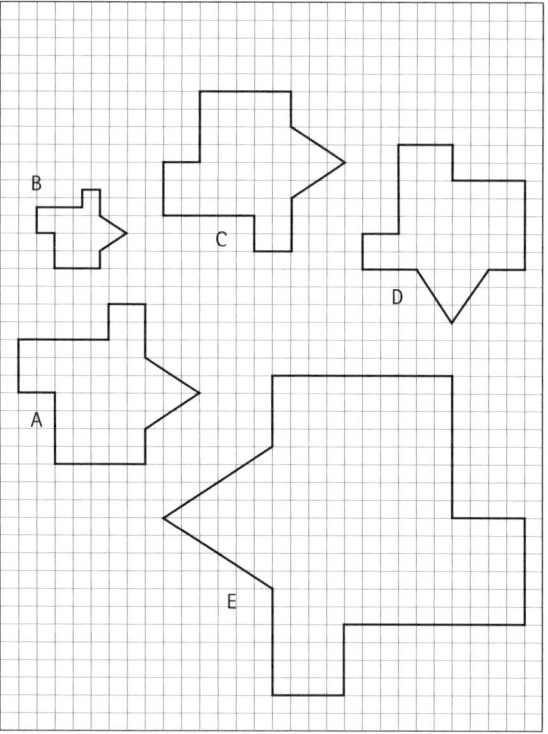

Fig. 1.24. Five similar shapes, with A, C, and D congruent

Among transformations, isometries and dilations are special because so much of our reasoning about figures depends on congruence or similarity. But artists, scientists, and mathematicians must also reason about *other* ways in which visual images do or do not change. Reflect 1.16 asks you to pause and think about one of these other transformations.

Reflect 1.16

Can you think of a transformation that is neither an isometry nor a dilation?

Explain why your transformation is *not* an isometry or a dilation.

Not every transformation is an isometry or a dilation; that is, they do not all result in similar shapes. Consider your response to Reflect 1.16 in relation to the transformation shown in figure 1.25. A "stretch" is an example of a transformation that does not result in a similar shape.

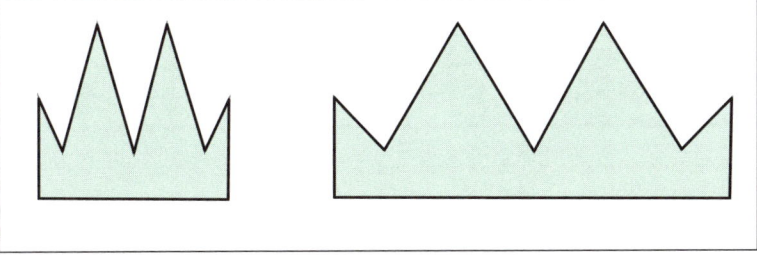

Fig. 1.25. A transformation resulting in a non-similar shape

The ability to describe, use, and visualize the effects of manipulating shapes and space is important in the sciences and the arts, for design and construction, and even for arranging furniture. Such competencies also provide a foundation for understanding number and arithmetic (for example, part-whole relationships in whole number arithmetic and fractions) and other areas of mathematics.

Recognizing invariant properties of transformations

Essential Understanding 3*b*. *Under each transformation, certain properties are invariant.*

The ability to see how shapes are composed—to see how smaller component shapes fit together to make the whole—is learned over time. The rectangular array in figure 1.26 might be used as a model for the introduction of multiplication or in explaining the meaning of area. Even such a "simple" array is not "seen" by young children in the way that we see it.

For a comparison
of the area model
with other models of
multiplication, see
*Developing Essential
Understanding
of Multiplication
and Division
for Teaching
Mathematics in
Grades 3–5* (Otto
et al. 2011).

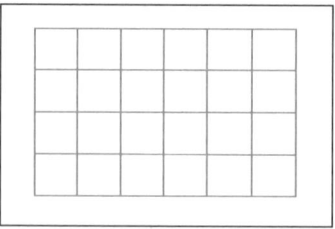

Fig. 1.26. An area model as a rectangle composed of unit squares

Adults are often surprised to see the drawings of early elementary students who believe they are copying this array but do not yet recognize the underlying rectangular structure, as illustrated by the students' work in figure 1.27.

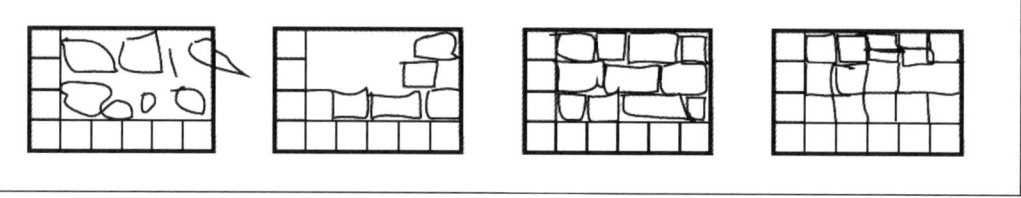

Fig. 1.27. Students' drawings of a 4 × 6 rectangular array

The children who drew the last two examples seem to recognize that the small tiles should completely cover the large rectangular region, but they do not yet appear to be thinking in terms of a clear row-and-column arrangement that ensures exactly four rows with six square tiles in each (or, equivalently, exactly six columns with four square tiles in each). In figure 1.28, the drawing is more accurate, but the student still appears to be drawing individual tiles, not vertical lines that subdivide the space into columns.

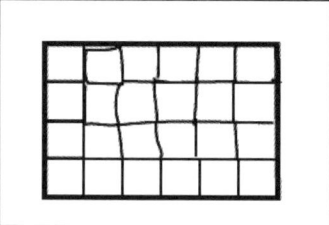

Fig. 1.28. A more accurate copying of a 4 × 6 array

Arrangements such as that shown in figure 1.29 are sometimes used in pattern-finding activities that are intended for older students and pose questions about how many blocks are in towers of various heights. Faced with this slightly more complex structure, even high school students struggle, and often fail, to draw models that correctly represent the number and arrangement of blocks.

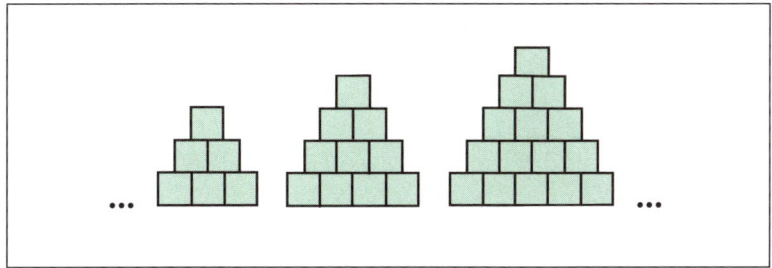

Fig. 1.29. A growing pattern of blocks for a pattern-finding activity

People use the geometry underlying the structuring of two-dimensional space implicitly and informally. They can *recognize* the diagrams in figures 1.26 and 1.29 as rectangles or "pyramid-towers," but their understanding needs to become explicit to support mathematical work. Understanding the row-and-column structure of a rectangular array is one basic idea. Full understanding also involves shape composition. That is, students need to learn to see each row simultaneously as 6 squares and as 1 row—a row *composed of* 6 squares. The row is a unit, just as each square is a unit: the row is a "unit of units." Similarly, the array can be viewed as a whole, 1 array, but also as composed of 4 rows—a unit of units of units. A view of a rectangle as "rows of columns" or "columns of rows" gives meaning to "length times width" or "width times length."

The connections between this geometric idea and ideas about number are not limited to multiplication. Its connection with place-value notation is also striking. The number 273 is composed of 2 of one kind of unit (hundreds), 7 of another kind of unit (tens), and 3 of a third kind (ones), but *each of those units* is also expressible in terms of the smaller ones. The 100 is simultaneously a unit (1 hundred) and a composite of units (10 tens), which are themselves composed of units (each ten is 10 ones).

Once students see *why* "length times width" is a reliable formula for the area of a rectangle (with whole number side lengths), they can compute the areas of other regions that can be *decomposed*, or *dissected*, into or from rectangles, as illustrated by the examples in figure 1.30.

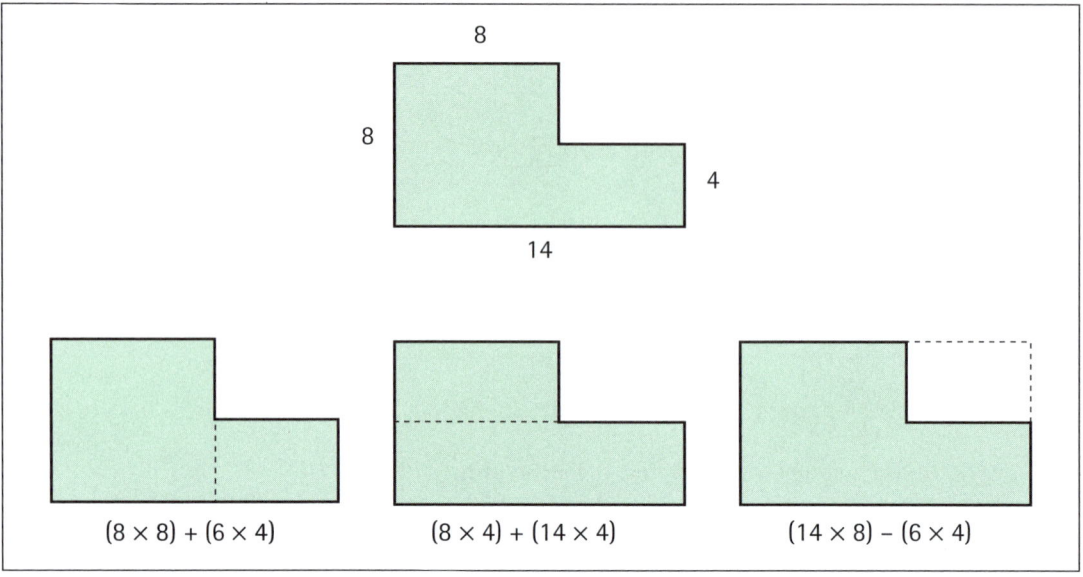

Fig. 1.30. Computing area by decomposition into rectangles

As another example of how composition and decomposition of shapes can serve in measuring, consider the common formula for the area of a triangle, $A = \frac{1}{2}bh$. Again, shape composition and decomposition can help bring spatial meaning to computations and formulas, as illustrated by the two strategies for finding the area of a triangle shown in figure 1.31.

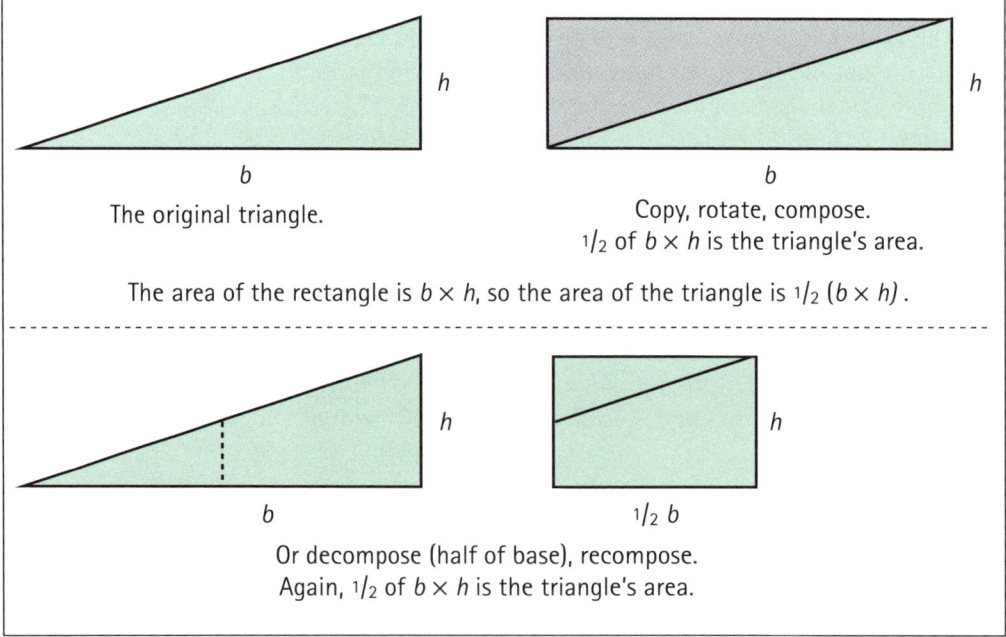

Fig. 1.31. Two ways of using composition and decomposition of a triangle to find its area

Transformations affect some properties of a figure but leave others unchanged. Recall that isometries—including reflections, rotations, and translations—preserve measures of the geometric figures but may change their orientations or positions. The illustrations in figure 1.32 are valid ways of thinking about the area of a parallelogram *because* changing the position of a figure does not change its area. So, *decomposing* the parallelogram and rearranging and reassembling its parts to make a rectangle produces a new shape altogether but leaves the area invariant. In fact, using all the parts with no overlap allows us to make an even stronger statement: *Any* polygon can be dissected (decomposed and recomposed) into any other that has the same area!

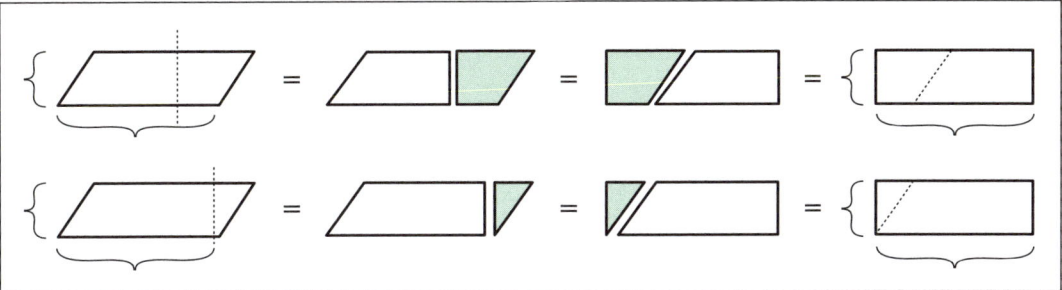

Fig. 1.32. Two ways of decomposing a parallelogram and rearranging the parts in a rectangle of the same area

Although it may seem obvious, it is important to note that such invariance is also important in one dimension. For example, reflecting, rotating, or translating a line segment does not change its length. When we choose a unit and repeatedly move it to span the linear extent of an object, the position of the unit changes, but its length remains invariant. The number of units that we have used in the iteration (without any overlap of units) is the object's *length* in terms of that unit. Further, a ruler can be perceived as a record of such iterations of a unit, and a number line can be viewed as an idealized ruler.

Dilations—magnifications or reductions (as on a copier set to enlarge or reduce)—do change measurements, of course. However, they also leave certain properties invariant. For example, we see that angle measures remain unchanged when we examine them with a magnifying glass.

Measuring Geometric Attributes: Big Idea 4

Big Idea 4. *One way to analyze and describe geometric objects, relationships among them, or the spaces that they occupy is to quantify—measure or count—one or more of their attributes.*

Geometry means "earth measure." One part of geometry is the study of geometric measurements, such as lengths, areas, volumes, and angles. Four essential understandings support Big Idea 4, which focuses on measuring attributes of objects.

Assigning a number to an attribute

Essential Understanding 4a. *Measurement can specify "how much" by assigning a number to such attributes as length, area, volume, and angle.*

We measure to determine the size, amount, or degree of something. Measurement is the process of assigning a number to the magnitude of some attribute of an object, such as its length, in relation to a unit. Measurable attributes are *continuous* quantities. In contrast, much of early mathematics involves *discrete* quantities. A discrete quantity is a collection of separate objects whose number can be determined exactly by counting with whole numbers. For example, we can count 4 apples exactly—that is a discrete quantity. We can add those 4 apples to 5 more apples and know that the result is exactly 9 apples. However, the *weight* of those apples is a continuous quantity, and scientific measurement with tools can give us only an approximate measure for this quantity. Whether we measure to the nearest pound or the nearest one-thousandth of a pound, there is always some error because measurement involves continuous quantities—amounts that can always be decomposed into smaller amounts.

One important kind of measurement is of the size of objects, but the word *size* lacks necessary precision. We might speak of size in at least three different ways. The size might be the length of the object (the extent of a one-dimensional object such as a line segment), the area (the interior region of a two-dimensional object such as a triangle), or the volume (the interior space of a three-dimensional object such as a cube). The "size" of a glass of milk is completely ambiguous: it might refer to the amount of milk (essentially the volume of milk) or its height in the glass. Without clarifying what is meant, we might reasonably argue about the "size" of the milk in two glasses that happen to have different diameters. One glass of milk could have more height but less milk. Which is the bigger "size"?

So we need to specify, but once we have done so, measurement is the process of assigning a number to a given continuous quantity by comparing it to an agreed-on *unit* of the quantity. That is, we determine how many units of the quantity make up the given quantity. To measure *length*—for example, the length of a room—we choose a unit of *length* (in feet or meters or whatever we like) and find how many of those units are needed to make up the length of the room. To measure *area*, we choose a unit of *area*, typically a square whose sides are one unit of length in whatever system of length units we like. And to measure *volume*, we choose a unit of *volume*, typically a cube whose edge is one unit long. In all three cases, we then find out how many of the units are needed to "match" the object we're measuring. We often need to subdivide the unit (feet into inches, or meters into centimeters) to achieve the precision that we need for the measurement.

Length can be viewed as a characteristic of an object found by quantifying how far it is between what we take to be the endpoints of the object. "Distance" is often used similarly to quantify how far it is between any two points in space. Measuring length or distance consists of two aspects, identifying a unit of measure and *subdividing* (mentally or physically) the given length or distance by that unit, by placing the unit end to end (*iterating* it) alongside the object. In some cases, like measuring the diameter of the earth, we can't do that directly, and so we depend on *calculating* the measurement from some other measurement that we can make. But even in those cases, the *idea* of the measurement is the same. When we describe the diameter of the earth as roughly 7900 miles, we are saying that it would take approximately 7900 of those mile units, end to end, to cover that distance.

Many ideas are included in the seemingly simple process of measuring length. We must understand that the unit of length is *repeated*, and each unit is the same length as each other unit unless, at the end, we need a common subdivision of that unit to add precision to our measurement. Thus, we might measure a child's height as 3 feet and 5 inches, but reporting that height as 1 yard, 1 decimeter, and 1 inch would seem strange and arbitrary because it would be inconsistent with the *idea* of choosing and sticking with a measurement unit. To make sense of measurements, we must be able to do more than just produce them. We must, for example, understand that if you are taller than me, and she is taller than you, then she is taller than me. This, of course, is a general truth about real numbers, not just numbers used to report measurements.

Indirect measurement is part of Essential Understanding 4*b*, the discussion of which treats it in greater detail.

Measuring length also involves understanding and applying the idea of the origin, or zero point, of a measure. On a ruler that has been marked off by iterating a unit of any size, any point can serve as the origin. We generally teach children to "measure from the end of the ruler," but the distance between 0 and 5 is the same as between 2 and 7, or even 31 and 36, so *any* point can serve as the "origin" as long as we then report our distances as distance *from it*.

Comparing and computing

Essential Understanding 4b. *Some quantities can be compared or measured directly, others can be measured indirectly, and the measurements of some objects are computed from other measurements.*

People often stand back-to-back to compare their heights. This process gives a direct comparison of a continuous quantity: length. Pouring water from one container that is initially full to another container that is initially empty can involve a similar direct comparison of the two containers' capacities. Alternatively, we can use a ruler to measure people's height directly, and a graduated cylinder to measure the containers' capacity.

Some quantities are difficult or even impossible to compare or measure directly. Indirect measurement may be needed because typical techniques of direct measurement are impractical. For example, to measure the height of a tree with a ruler or tape measure is difficult at best in many cases.

We might measure the height of a tree indirectly in several ways. We might stand back, hold a stick at arm's length, mark the sighted length of the tree from the base to the top, as illustrated in figure 1.33a. Then, as shown in figure 1.33b, we could rotate the stick to indicate approximately where on the ground the same length would fall. We could then measure this length, which is the distance from *A* to *C* in figure 1.33b.

Another way to measure a tree's height involves the use of similar triangles and proportions. For example, as illustrated in figure 1.34, one could sight along a known height, in this case a "curve ahead" highway sign, and use proportions to calculate the height of the tree.

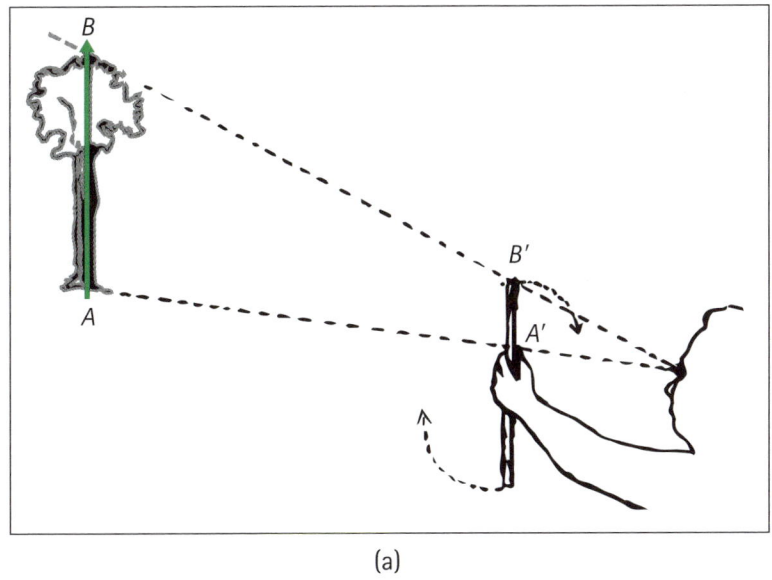

(a)

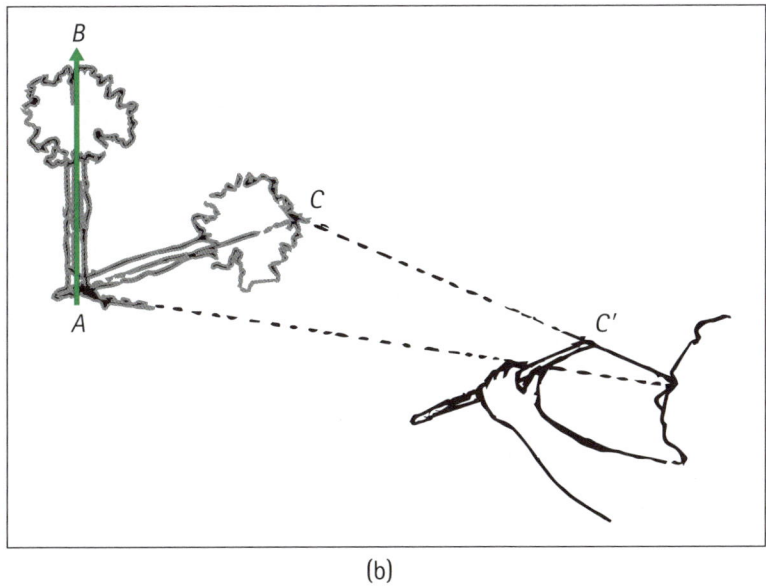

(b)

Fig. 1.33. Measuring height of tree using distance along ground

See *Developing Essential Understanding of Ratios, Proportions, and Proportional Reasoning for Teaching Mathematics in Grades 6–8* (Lobato and Ellis 2010) for an extended discussion of speed and other indirect measurement settings in terms of ratio and rate.

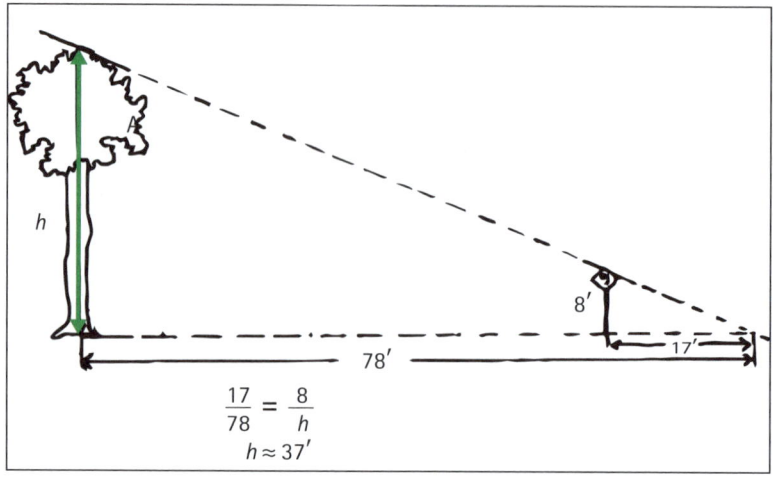

$$\frac{17}{78} = \frac{8}{h}$$
$$h \approx 37'$$

Fig. 1.34. Measuring height of a tree by using similar triangles and ratios

Indirect measurements can help solve problems for which making direct measurements is impossible or impractical, as in the case of finding the height of a tall tree. Finding the diameter of a planet through indirect measurement is more challenging! Reflect 1.17 gives you a chance to think about this task.

Reflect 1.17

How might you find the diameter of a planet?

Finally, some measurements are computed from a combination of direct or indirect measures. Speed is a simple example, calculated as distance traveled divided by the time of travel.

Relating unit size to the number of units

Essential Understanding 4c. *Measurement can be performed with a variety of units. The size of the unit and the number of units in the measure are inversely related to each other.*

Another understanding that proficiency in measurement requires is the proportionality of measurements, including the inverse relationship that exists between the size of a unit and the number of units in a given measure. Consider the two rows of matches in figure 1.35.

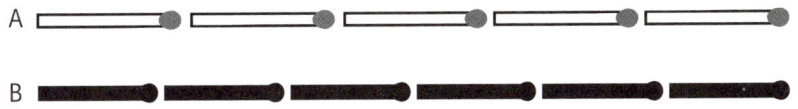

Fig. 1.35. Measurement with different units

Although from the adult perspective the lengths of the rows are the same, many children will argue that the row with shorter matches is longer because it has more matches. Even to adults, 254 centimeters often "sounds longer" than 100 inches. One needs to understand the relationship between the size of the units and the number of units to understand measurement situations. Measurement is not just about counting the units.

Composing and decomposing

Essential Understanding 4d. *Objects can be decomposed and composed to facilitate their measurement.*

The distance around a rectangle—its *perimeter*—can be easily found by adding the lengths of the rectangle's sides. This illustrates a basic idea: Objects can be taken apart and put together in various ways to help determine their measurements. The decomposition of polygons to determine their areas, highlighted in the discussion of Essential Understanding 3*b*, is an example of how decomposing and composing geometric figures aids in their measurement.

Essential Understanding 3*b*

Under each transformation, certain properties are invariant.

Conclusion

In this chapter, four big ideas related to geometry provide a foundation for teachers' understanding and support their efforts to help young children to make sense of the world around them. Each of the big ideas focuses on essential mathematics that helps teachers formalize their thinking about the ways in which children are "seeing" their world. Big Idea 1 addresses classification, and our discussion has pointed out how children see shapes and relationships across groups of shapes. Attention to geometry as a means of structuring space and specifying locations is the focus of Big Idea 2, which provides another means to understand how children see the world around them. A third way in which children see and make sense of their world is by noticing what changes and what does not change, and this is the intuitive basis of Big Idea 3. Following from visualization of shapes and their associated characteristics is the need to measure or count particular attributes of the shapes to see

Big Idea 1

A classification scheme specifies for a space or the objects within it the properties that are relevant to particular goals and intentions.

 **Big Idea 2**

Geometry allows us to structure spaces and specify locations within them.

 **Big Idea 3**

We gain insight and understanding of spaces and the objects within them by noting what does and does not change as we transform these spaces and objects in various ways.

 **Big Idea 4**

One way to analyze and describe geometric objects, relationships among them, or the spaces that they occupy is to quantify—measure or count—one or more of their attributes.

the world analytically. This need underlies Big Idea 4. All of these big ideas work in tandem as teachers help children navigate the many complexities of their world.

In the next chapter, we look back at early spatial awareness and ahead at how the visualizations articulated in the big ideas and essential understandings link to the school curriculum beyond grade 2. Geometric concepts and skills that teachers need and children encounter in later grades use the ideas in this chapter as a foundation for more complex and sophisticated thinking about and applications of geometry and measurement.

Chapter

2

Connections: Looking Back and Ahead in Learning

Geometry and measurement help us understand the space in which we live. Because they also connect closely with other areas of mathematics, teachers of children of *all* ages, regardless of the mathematical content that they are actually teaching, must understand some key ideas in geometry and measurement. This comes as no surprise.

Seeing Connections with Geometry at Other Age Levels

Each of the big ideas and essential understandings elaborated in chapter 1 helps teachers understand how children's geometric thinking develops across the grade bands. It is useful to begin by looking back at the earliest years.

Locating and visualizing: The earliest geometry

From birth, children possess remarkable competencies in observing and moving within their spatial world. Infants can focus their eyes on objects, and soon they begin to follow moving objects with their eyes. Toddlers use geometric information about the overall shape of their environment to solve location tasks; this is the intuitive basis of Big Idea 2. For example, figure 2.1 shows a baby and a mother, as seen from above, with the baby in a highchair and × marking a familiar toy in the baby's field of vision in the space between them.

 Big Idea 2

Geometry allows us to structure spaces and specify locations within them.

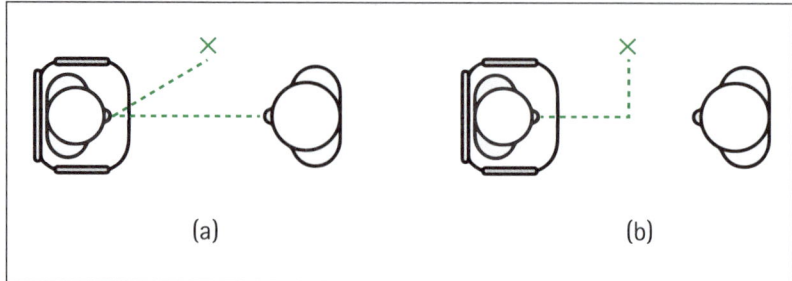

Fig. 2.1. As seen from above, a baby in a highchair opposite a mother, with × marking the location of a toy between them

The dotted lines in figure 2.1a indicate the baby's perception of the toy as about 20 degrees left of the mother and slightly closer than she is—without, of course, knowing anything about degrees or measuring distances. Yet, the baby does have a sense of orientation in space and of comparative distance.

The baby certainly does not locate the toy by thinking in *x*- and *y*-coordinates: "so much distance ahead and then so much distance directly to the left," as suggested by the dotted lines in figure 2.1b. A child thinks much more naturally in terms of direction and distance to a goal (polar coordinates) than of distance left-right and forward-backward (Cartesian coordinates).

The "geometry in action" that the baby does in locating the toy lays an intuitive foundation for a variety of mathematical topics, such as work with paths and polar coordinates that he or she will encounter much later. Very young children start to develop the foundation for these topics as they remember a location or route to a location through a pattern of movements associated with a goal.

Later, they learn entire paths. They remember locations as distance and direction of their own movements and landmarks found along that path. Crawling from the kitchen to the playroom, the child learns the location of the door, the turn, the mirror that marks the playroom's door. They build these competencies from an internal reference system—the reference is the self, moving through space. They lay the developmental foundation for understanding the concepts of geometric paths, straight paths, paths with bends (angles), distance and length, and, eventually, differential geometry (the study of the geometry of curves and surfaces).

Toddlers are also building experience with externally based reference systems. These systems and the experiences gained by using them in these early years support understanding of coordinate systems. The beginnings are spatial relationships within and between environmental structures and landmarks. Such landmarks are initially objects that are familiar and important. For example, the child might remember last seeing a toy under a couch against a certain wall and, moreover, might recall that the toy was closer to the end of the couch that is by the door. Such competencies develop into "mental maps" because they build knowledge of locations from distances and spatial relationships among environmental landmarks, structuring space as captured in Big Idea 2.

Children later may recall that the sand shovel was buried about halfway out from the wall of the garage and "about this far" from the edge on the street side of the sandbox, a remembered location that shows the development of precision central to Essential Understanding 2*b*. This is an early use of intuitive geometry that will later be articulated as Cartesian coordinate systems, as described in the discussion of that essential understanding.

At first, children naturally see most objects in their environment from many points of view, and they code all those distinct images as the same object without attending very much to how the different viewpoints affect the image. Their frequent early confusion of **p**, **b**, **d**, **q** (reflections and rotations of one another), as discussed in chapter 1, illustrates this point, which applies equally well to objects and pictures. Early geometric learning involves beginning to *notice* that the position of an object can change the way it looks. Children know that it is the same object, but they now also notice how the appearance can change. For example, the rim of a paper cup can seem either circular or elliptical or even appear to be a straight line, depending on how the cup is tilted. It looks circular when they stare down on it, straight into the cup, and looks like a straight line when they see the cup from the side, with their eyes level with the top.

The opposite occurs, too. Because we have special neurons devoted to recognizing *vertical* lines (presumably to help us remain upright ourselves), we privilege features that are "on top and bottom" or that suggest vertical lines. As a result, young children *see* the two shapes in figure 2.2 quite differently. They are most attuned to the vertical and horizontal lines of the first, and most attuned to the vertically and horizontally aligned *corners* of the second. A young child will actually *draw* the second figure not as four lines, but as four corners, not necessarily even making straight-line connections between those corners.

Big Idea 2

Geometry allows us to structure spaces and specify locations within them.

Essential Understanding 2*b*

Geometry and measurement can precisely specify directions, routes, and locations in the world—for example, navigation paths and spatial relations—with precision. Given a reference point and an orientation, we can label position with numbers.

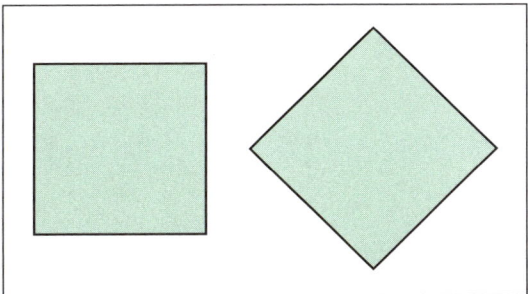

Fig. 2.2. A square in different orientations

It is no surprise that they, and we, refer to the second shape as a "diamond" or at least as a "turned square" (or "rotated square"). Even after we know that the two are "the same shape," they don't look the same. These two squares are, by the way, the same size, but often people see the "diamond" as larger. So, sometimes we need to notice that things really are different, although we "see" them naturally as the same; and sometimes we need to notice that things really are the same, although we see them as different. Both require learning.

Further, children learn spatial and geometric vocabulary—terms that signify position, such as *in, on, under, up, down, beside, between, in front of, behind,* and later, *right* and *left.* This vocabulary is the linguistic basis for connecting children's early, intuitive ideas with the refinements and extensions that we know as mathematics. Intuitions become more precise models of everyday situations when we use mathematical ideas of number and shape, mathematical actions such as measuring or transforming shapes, and structural relationships among these ideas and actions. Mathematics involves both systematizing (refining, extending, and relating) these ideas and actions, and using the resulting models to solve problems. Learning mathematical language and using it are essential to this process of *mathematizing.*

Bringing together ideas related to location and transforming shapes, children can mathematize their experiences with navigation and spatial relationships as they use and create simple models and maps. Block building (see fig. 2.3), including making models and maps of the classroom or playground, capitalizes on many of these experiences and has meaning when viewed by teachers who have grasped Essential Understanding 2*b*. Building with blocks also relates to students' later experiences with coordinate systems and the spatial structuring that underlie the measurement of area and other topics both in and out of geometry. For example, making a floor for a building with square blocks can be the beginning of spatial structuring. The ability to organize objects into rows or columns, or into distances from axes, can begin with such activity and is meaningful to teachers with a firm grasp of Essential Understanding 2*a*.

➡ Essential
Understanding 2*b*

Geometry and measurement can precisely specify directions, routes, and locations in the world—for example, navigation paths and spatial relations— with precision. Given a reference point and an orientation, we can label position with numbers.

Fig. 2.3. A child building with blocks

Essential
Understanding 2a

To describe a location, we must provide a reference point (an origin) and independent pieces of information (often called coordinates) indicating distance and direction from that point.

Ideas about geometric shapes have their beginnings in early geometry. Shape is central to young children's understanding of the world. Children's attention to shape goes far beyond their learning names of familiar shapes such as circles, although that is important. When children learn new words for objects (such as "salt shaker"), the shape of that object is the main feature that they use, rather than the color, size, sound, or other attributes (Smith et al. 2001).

The shape of almost every object is a combination of "basic" shapes. In fact, that is what we mean when we *call* a shape "basic": it is the *basis* of many familiar but more complicated shapes. Think of the common simple representations of people with circular heads and rectangles for many other body parts, or the house represented as a triangle on top of a square. Such idealizations of shape are not restricted to children's immature ways of drawing. Artists are often taught to "see" complex forms geometrically and to lay out the forms with purely geometric guidelines, as in figure 2.4, before filling in the details that make the artwork look more "real."

Fig. 2.4. An example of an artist's use of simple geometry figures

"Basic" shapes, just like complex forms, can also be decomposed, and children gain a head start in understanding some of these decompositions when they combine, for example, two blocks that are triangular prisms to make a prism that is square or rectangular, or a larger triangular prism. As when children lay down an array of blocks to make a floor, such shape compositions build the conceptual foundation for understanding geometric compositions, length, area, volume, and coordinate systems.

Building to geometry in grades 3–5

The geometry learned during the early childhood years forms a critical foundation for geometry and measurement in grades 3–5. Early learning of shape becomes more systematic, with students *classifying* shapes explicitly on the basis of properties, and *drawing inferences* about the shapes on that basis (indicative of early work related to Big Idea 1). For example, students can now use a rectangle's four right angles to make inferences about the relationships among its sides, pairs of which must be parallel and the same length. Hierarchical inclusion—which involves recognizing that some classifications are subsets of other classifications—also begins more explicitly. For instance, students become aware that squares are special cases of both rectangles and rhombi. They can compare different ways to define a shape category (for example, defining a rectangle as a quadrilateral with all right angles or as a parallelogram with a right angle).

Children in grades 3–5 build on their early work with length measure to address precision and subdivisions of units. Their knowledge of length measurements, origins, units, and distances becomes the foundation for their learning of a Cartesian coordinate system, which is one way to structure space and specify locations (Big Idea 2).

Big Idea 1

A classification scheme specifies for a space or the objects within it the properties that are relevant to particular goals and intentions.

Big Idea 2

Geometry allows us to structure spaces and specify locations within them..

Early work with sliding, flipping, and turning objects provides students with experiences that they later consider systematically as the geometric motions of translations, reflections, and rotations, respectively. Each of these transformations has particular characteristics and specific geometric attributes that it does or does not preserve. To describe the characteristics or attributes of a rotation, for example, one must specify a point around which the rotation occurs—the *center of rotation*—and an amount of rotation (an angle, taken to be counterclockwise by convention but typically specified in elementary school with *both* an angle *and* a direction clockwise or counterclockwise, for clarity). In the later grades, students study such characteristics explicitly. Further, the implications of these transformations—what does or does not change under a particular transformation—are also made explicit and studied. The mathematics of these transformations is discussed in chapter 1 (Big Idea 3), but most are developed in a school curriculum for students in grades 3 or above.

Students in the intermediate grades extend earlier shape composition to reflect on spatial structuring, area, and volume. The next section discusses these measurement topics.

We gain insight and understanding of spaces and the objects within them by noting what does and does not change as we transform these spaces and objects in various ways.

Making Connections with Other Topics

The concepts that are specific to geometry and measurement are complemented by essential mathematical habits of mind, many of which are shared by other mathematical domains, illustrating both the coherence of mathematics and the centrality—and therefore the usefulness—of these habits of mind for educators and students. Connections between geometry and the rest of mathematics both illustrate the elegance of mathematics and help students see that one domain can imbue another with new meaning.

Using number lines

Spatial, geometric, and measurement competencies connect directly with one of the basic models used in number and arithmetic: the number line. For mathematicians, just as for school children, the number line is a way of visualizing numbers both as locations and as distances between those locations. Each point on the number line is uniquely identified with a number, just as a house is uniquely identified with its address along a street. Houses generally use only whole numbers as addresses; likewise, number lines for the youngest children typically show only whole numbers.

By contrast, a mathematician's concept of a number line includes *all real numbers*—positive ones to the right of zero, and negative ones to the left, and all fractions and all other numbers as well, filling in the spaces. The number line, as children first see it in school, is generally shown as a horizontal line, with a point designated as zero and equally spaced points labeled 1, 2, 3, 4, ... representing the whole numbers. Pictures of the number line come, as all pictures do, in various sizes. But whether the space between consecutive numbers is an inch or a centimeter, as it might be on a ruler, or a much larger space, as wall models usually require, *that distance* between two consecutive numbers on *that line* is considered to be "the unit" for that number line. This allows us to say that the distance between 7 and 10 is 3. The line segment from 0 to 1 is the conventional, or standard, example of the unit segment, and the number 1 is also called the unit, with numbers serving both as locations and as distances. Once we have determined this, all the whole numbers are fixed on the line.

Rulers are portable, physical representations of finite parts of number lines, so they can be readily used to find distances. Each ruler has its own unit—inches, centimeters, or, for special purposes, other units (or no *named* unit at all, as on wall number lines, with only the regular spacing serving as the unit). Thus, when we report a length (distance) that we measure with a ruler, we must also specify the unit of that particular ruler: the distance between the 7 and the 10 on an *inch*-ruler is 3 *inches*.

The number $1/2$ is exactly halfway between 0 and 1. We call the distance from 0 to $1/2$ "one-half." We see that exactly three of those distances to the right of 0 is a point exactly halfway between 1 and 2, which can be called "three halves" ($3/_2$) because it is three *halves* from 0. We can also call that point "one and one-half" ($1 1/_2$) because it is 1 *and* $1/2$ units from 0. (This is an example of number serving as location and distance.) Rulers marked in inches often subdivide each unit into halves, quarters, and even smaller subunits. We can subdivide units on the number line any way we need, to find thirds, or eighths, or tenths, or seventeenths.

The number line clearly connects directly with linear measurement. It represents numbers—whole or rational or irrational (numbers that can't be expressed as 10ths or 17ths, or any other *n*ths)—as the length or distance from 0 to that number. Thus, both measurement and associated number line models can serve as tools for mathematics: number (including fractions and decimals), number comparison (any number "to the right" of another number is larger than that number), arithmetic (subtraction uses the distance between numbers as a way to compare them), and estimation (finding numbers "in the neighborhood"—that is, not too distant).

Number lines are especially important for understanding rational numbers, including fractions and their decimal representations. They can be used to illustrate the relationship between fractions and whole numbers and to demonstrate that fractions *are* numbers,

For more details about connections between the number line and measurement, see *Developing Essential Understanding of Number and Numeration for Teaching Mathematics in Prekindergarten–Grade 2* (Dougherty et al. 2010).

that fractions include numbers greater than 1, that fractions can be added on a number line model in a way that closely mirrors adding whole numbers on a number line, and so forth. In all these ways, measurement can support building "mental number lines," which research suggests play a critical role in understanding mathematics (e.g., Elia, Gagatsis, and Demetriou 2007; Geary et al. 2008; Ramani and Siegler 2008; Rodriguez, Parmar, and Singer 2001; Vanbinst, Ghesquiere, and Smedt 2012).

Composing, decomposing, and unitizing

The composition and decomposition of shapes are core processes in geometry. Such processes, and the conservation of area, are useful conceptual tools for solving problems involving tessellations, area, and so on. However, these tools are also useful for understanding number. For instance, although the number line serves best to show fractions *as numbers*, fractions also describe amounts *of* something, and the decomposition of a geometric whole into parts of equal area can serve as a useful model of that use and aspect of fractions. Consider an example. Suppose that we choose the area of the equilateral triangle in figure 2.5 as the unit with which to measure the area of the larger shape. The area measure is then 18 units. It takes 18 of the units to cover the region.

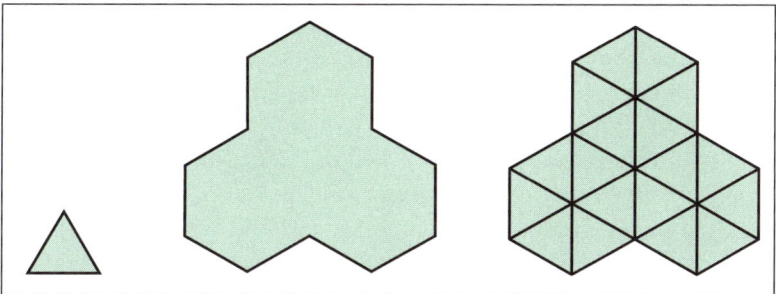

Fig. 2.5. An equilateral triangle used as the area unit to measure area of a figure

If we choose, instead, to measure the same region with a unit that is exactly 6 times the area of the original unit—that is, with the area of a regular hexagon in place of the equilateral triangle—we will now use $1/_6$ as many units as before. We will now report the area as 3 units. Figure 2.6 provides examples for $1/_3$ $1/_9$, and $1/_{18}$. This experience exactly parallels the experience with different size units for measuring length and helps establish a general truth about measurement, not just a specific kind of measurement.

In *Developing Essential Understanding of Rational Numbers for Teaching Mathematics in Grades 3–5*, Barnett-Clarke and colleagues (2010) use number line, unit, and iteration to interpret rational numbers as measures.

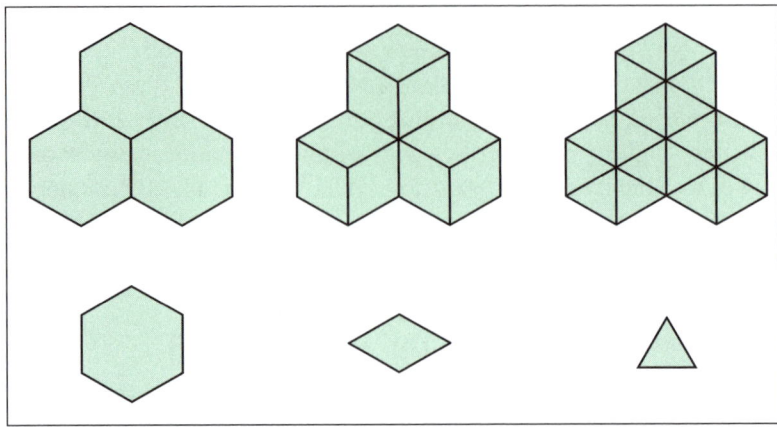

Fig. 2.6. Finding the area of same shape with different units

If the area unit is the area of the hexagon, the area is 3 units. If the unit is $1/_3$ that size—the area of the rhombus—the area is 3 times as many units, or 9 units. If we change units to use units of one-third the size, we will need three times as many of them to measure the area: $1/_3$ is the multiplicative inverse of 3. If we choose instead to measure the area with a unit that is $1/_2$ the size of the rhombic area unit, this time using the area of the triangle, we will need two times as many of them to cover the region: $1/_2$ is the multiplicative inverse of 2. Working with the area of the triangle as the unit, we will report the area as 18 units. The *number* of units that we report—the area measure—depends on the *size* of the units that we use, and the relationship is an inverse one under multiplication: the smaller the unit, the larger the measure.

The inverse relationship between the size of the unit and the number of units that compose a given shape or quantity is thus modeled in geometric shape composition and in area measure-ment. It also mirrors the inverse relationship between the size of the counting numbers (increasing as 1, 2, 3, 4, …) and unit fractions (decreasing as $1/_2$, $1/_3$, $1/_4$, …).

Further, all numbers are structured by composition related to a specific unit. The decimal place-value system groups by tens. Base-ten blocks are designed to embody this structure. If we choose the length of one edge of a small cube to represent "one," then the length of the rod formed by the lengths of the edges of ten cubes represents "ten," as in figure 2.7. This is similar to the composition of shapes that are combined to form a larger shape that is concep-tualized as a new shape (a rectangle, say) as well as a composite unit of smaller shapes (say, squares). This reflects the importance of viewing a number flexibly as "10 tens" and "1 hundred."

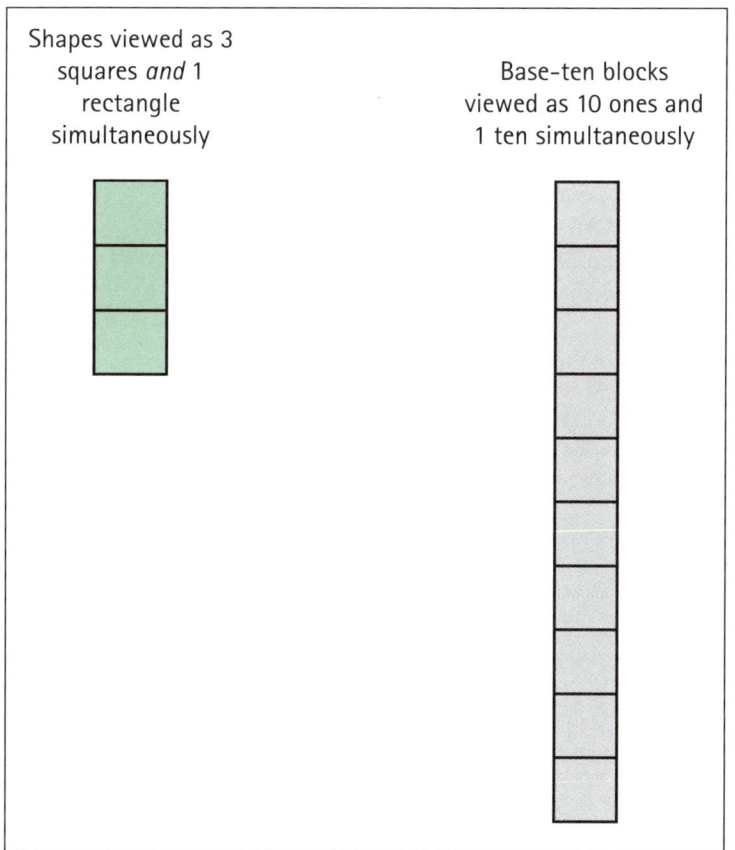

Shapes viewed as 3 squares *and* 1 rectangle simultaneously

Base-ten blocks viewed as 10 ones and 1 ten simultaneously

See *Developing Essential Understanding of Number and Numeration for Teaching Mathematics in Prekindergarten–Grade 2* (Dougherty et al. 2010) for an extended discussion of unit and place value.

Fig. 2.7. Viewing shapes as a composition of units

Having a firm grasp of composition and decomposition in geometry and measurement broadens students' understanding of parts and wholes and provides concrete models for numerical operations that combine, separate, or compare numbers.

Working with spatial structures

A particularly important example of composition and unitizing is in spatial structuring. One way to view the area of a rectangle is as partitioned into identical square units organized in rows and columns, as illustrated in chapter 1 (see fig. 1.26). Such decompositions, organized in rows and columns, lend a mathematical interpretation to the formula *Area = length × width*. Each row contains the same number of square units, and the number of rows in the rectangle equals the number of squares in a column.

Another connection between geometry and arithmetic involves rectangles and area. The commutative property of multiplication (by which, for example, 5 × 3 = 3 × 5) connects with the notion that one can group the squares in the columns together into units, and then the number of squares in a row equals the number of those units. Alternatively, one could consider that a geometric motion, a 90-degree rotation of the rectangle, "switches" the rows and columns.

The rotation of a rectangle is also an application of conservation of area. As is true of any shape, the area of a rectangle does not change with a rotation. However, young children may think that if a rectangle is rotated so that it is "taller" than the original rectangle (see fig. 2.8), it will have more area.

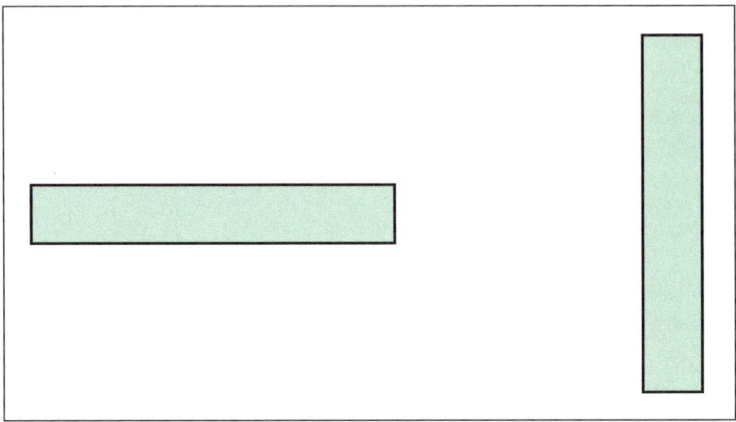

Fig. 2.8. A rectangle rotated to produce a "taller" rectangle

Area models for multiplication are important for another reason. Using arrays of discrete objects (for instance, 4 rows of pennies with 8 in each row) to model multiplication works well for whole numbers, but not in general for rational numbers. For example, 2.5 rows containing 5.1 pennies each makes no sense. Because area models based on measurement are continuous models and can be partitioned into units of any size, they lend themselves well to rational numbers and thus to the multiplication of fractions and decimals.

Just as the area of a rectangle tiled by unit squares can be measured by counting unit squares per row and multiplying by the number of rows—that is, multiplying rows by columns, height by width—the volume of a rectangular solid can be measured by counting the unit cubes in a layer (multiplying rows by columns in that layer) and multiplying by the number of layers. The formula *Volume = length × width × height* is explained by this spatial structure.

Thinking of a plane as an infinite array is a different spatial structuring. This structuring relates to a Cartesian coordinate system, in which an ordered pair of numbers identifies each location, or point, in the plane and indicates its distance from two coordinate axes.

For a discussion of the meaning of multiplication and properties of multiplication as illustrated in array models and other models of multiplication, see *Developing Essential Understanding of Multiplication and Division for Teaching Mathematics in Grade 3–5* (Otto et al. 2011).

Conclusion

The sophistication of geometric concepts required in later grades builds on the experiences that students have in the early grades. Understanding spatial relationships is one of the most important outcomes from children's experiences. Children learn to structure and understand space in new ways. It is through spatial relationships that children come to understand transformations of shapes in planes and space, discerning what is or is not changed by the transformations.

Measurement concepts are closely aligned with transformations as children consider iterations of units to form measurement tools. This leads to an understanding of how the size of the unit affects the measurement count.

Even though geometry is often not stressed in school mathematics in the same way that number and operations are, the topics centered on geometry and measurement clearly link to and support development across other mathematical areas. Teachers who understand how these topics support learning beyond geometry can use geometric concepts to support students in building stronger knowledge related to other topic areas.

Challenges: Learning, Teaching, and Assessing

Our examination of the big ideas of geometry and measurement has demonstrated that these topics go far beyond the notions of teaching shape names and showing how to use a ruler. Geometry and measurement are fertile fields for mathematical concepts, skills, and reasoning in and of themselves, as well as for connections with basic concepts and competencies in number, arithmetic, and other mathematical topics.

Supporting Spatial Understanding

Geometry and measurement require and can help develop general spatial skills. This idea is particularly important to understand because research has demonstrated that teaching spatial skills in isolation is usually unsuccessful (Hofmeister 1993). Integrated approaches that combine spatial thinking with other topics, especially geometry and geometric measurement, are more effective.

Teachers who understand the big ideas identified in chapter 1 and the connections among them begin by planning the school environment. Not only do they ensure that their classrooms include unit blocks, puzzles, and tangrams, but they also plan interesting layouts of objects inside and outside their classrooms. They provide incidental and planned experiences with landmarks and routes and initiate frequent discussion related to spatial relations, with supporting activities, including creating directions, throughout the school day (Clements and Sarama 2009; Sarama and Clements 2009).

By hiding an object somewhere in the classroom and then providing a clue, such as a photo taken of the hiding place from some unfamiliar position or angle or distance (perhaps very close up or from above), teachers offer a nice puzzle to young children. Alternatively, the clue can specify a starting place, a number of steps in a particular direction, a turn in one direction or another, and another set of steps. In devising clues of this sort, teachers draw on their awareness of Essential Understanding 2b.

These types of puzzles and games lead seamlessly into map making. In mathematics or other subjects, such as social studies, explicit discussions that connect real-world space and maps help children understand the mathematics of mapping. Such work should address the four mathematical questions identified in the discussion of Essential Understanding 2b:

- Which way? (direction)

- How far? (distance)

- Where? (location)

- What objects? (identification, using symbols)

To begin, children might work with blocks and toys to build maps of the classroom or playground. Students in the primary grades might measure (perhaps by pacing or using a trundle wheel on the playground), and translate that information (including the origin, direction, and distance) into a model in the classroom. Similarly, combining physical movement, paper-and-pencil work, and time on the computer can help with such mathematizing. As one example, children might translate a set of directions or a map onto a screen and then work in a digital environment such as Scratch or Logo to tell an animated character the movements to make to traverse a given route or trace a given shape on screen.

Locations on maps can include an introduction to coordinates. For the youngest children, a "tiny town" with a few labeled streets or avenues can be modeled on the floor with blocks and tape, as in figure 3.1. Children can place a house at B Street and 4th Avenue or give directions for someone to travel from that house to a school somewhere else on the grid—saying, for example, "Go 2 blocks north on B Street and then turn west and go 1 block."

Children learn to understand and eventually quantify what grid labels represent. To do so, they need to connect their counting acts with those quantities and labels. They need to structure grids mentally as two-dimensional spaces, demarcated and measured along each dimension (a type of spatial structuring). That is, they need to understand coordinates as a way to organize two-dimensional space by coordinating two perpendicular number lines. Every location is named by two measurements, one given by each of these two number lines. Essential Understanding 2b underlies the teacher's understanding of the mathematical importance of learning these things.

Real-world contexts such as maps of the school or surrounding area are helpful in teaching coordinates initially. So too are many other activities that build a foundation for spatial structuring. These

Essential Understanding 2b

Geometry and measurement can precisely specify directions, routes, and locations in the world—for example, navigation paths and spatial relations—with precision. Given a reference point and an orientation, we can label position with numbers.

Logo is a computer programming language that allows children to give directions such as "forward 10 steps, right turn, forward 5 steps" to an on-screen animated turtle to navigate a maze or draw geometric objects.

Scratch (http://scratch.mit.edu/) is similar to Logo and provides an interactive programming environment and online community in which children can create, view, and share projects that embody movement.

might include a variety of children's games, including the traditional favorite known as "memory" or "concentration," in which two copies of various shapes are printed on cards that the children shuffle and arrange facedown in a rectangular grid. The children then take turns to draw a pair of cards. If the two cards match, the child keeps the "match" and plays again. If the cards don't match, the child returns them to their original places in the grid and tries to remember what row and column they occupy to make matches later. Similarly, children might create shapes on geoboards by following the directions of a teacher, who tells them, for example, "Start on the nail in the second row from the bottom, in the fourth column from the left."

Fig. 3.1. A child working with a "tiny town"

Mathematical ideas should be clearly articulated throughout instruction at every level, with the contexts gradually fading and eventually giving way to more abstract consideration of the mathematics of coordinates (interspersed with other activities to maintain the connection with real-world contexts). Building well-defined mathematical concepts to support spatial structuring requires clear and unambiguous language from the teacher—what the Standards for Mathematical Practice (Common Core State Standards [CCSSI 2010]) refer to as "precision" (p. 7)—and what might be thought of as a kind of "mathematical hygiene." Casual phrases such as "over and up" may be clear enough if everyone can see the same object and the speaker is pointing, but they don't model the precision of language that to necessary to help children clarify their own communication. For example, does "over" mean "left" or "right" (or "east" or "west"), and does "up" mean "farther from the floor" or "higher on the page"? Similarly, the statement, "The x-axis is the bottom," may be true of a particular picture, but it does not generalize to a four-quadrant grid, where the x-axis remains the horizontal one by convention but is not at the bottom of the page. Further, such wording may hinder the *integration* of coordinates into a coordinate pair indicating a single location. The precision in language in these examples is similar to the precision in thinking that must develop to support classifications as described in Big Idea 1.

Creating maps and coordinate grids, interpreting those of others, and connecting maps and grids to real-world locations are tasks that not only develop children's competence in structuring space and specifying locations in space but also help teachers assess these competencies. Teachers who organize their thinking around the big ideas and essential understandings see such tasks as more useful educationally than isolated exercises in plotting points specified by their coordinates.

Big Idea 1

A classification scheme specifies for a space or the objects within it the properties that are relevant to particular goals and intentions.

Building Awareness of Shapes and Transformations

Experiences and instruction build students' knowledge of geometry. Teachers who provide a rich variety of examples and non-examples of geometric shapes help students form flexible and accurate categories and avoid common misconceptions (such as the notion that rectangles must have pairs of sides with different lengths or that something is not a triangle if it diverges from the pedestrian isosceles triangle with a horizontal base).

Such teachers also keep their vocabulary consistent, clear, and accurate. Young children often say that a triangle has three sides. That's completely true, but teachers who take that as an invitation to discussion often find that children accept sides that are not straight, regard a non-convex angle as "one side" (mistakenly saying that

is a 3-sided shape), or even count the vertices rather than the sides.

Recognizing these potential misconceptions, teachers can provide varied examples and non-examples to help children understand the *defining* attributes of shapes and understand that other attributes, such as color, position, and size, are not relevant to the classification of a shape (although position and size are certainly relevant in other mathematical contexts). The point, of course, is to help children acquire both the relevant features and new and useful language to specify clearly what they mean—not to hinder their progress by constantly bogging down in detail or forcing them to say more than is necessary for clear communication.

Teachers who help students develop knowledge of shapes benefit from Essential Understanding 1*a* as they include four features in their instruction (Clements and Sarama 2009; Sarama and Clements 2009):

1. Varied examples and non-examples

2. Discussions about shapes and their attributes

3. A wide variety of shape classes

4. A broad array of geometric tasks

Triangles of many shapes and sizes might be discussed, challenging children to describe *why* a certain shape is or is not a triangle. Other shape classes could be similarly explored and discussed. For example, given a set containing rhombi, parallelograms, and polygons with more than four sides, students could classify them into two subsets: "quadrilaterals" and either "not quadrilaterals" or "polygons with more than four sides." The classifications do not need to be specifically named, although that, too, is an option if the children would find the naming interesting. In addition, seeing shapes that are not polygons at all—because they are not closed, or because not all the "sides" are straight, or for some other reason—helps children create the classification correctly. Contrast makes things clearer! It also helps children look for and acquire words to describe the way or ways in which these figures "fail."

Tasks that promote reflection and discussion, such as building models of shapes from components, become reasons for wanting language that supports clear communication. For such construction tasks, children can choose from a variety of materials, such as sticks of varied lengths and chenille sticks (often called "pipe cleaners"), to create good examples of triangles, trapezoids, parallelograms, rectangles, squares, rhombi, arbitrary quadrilaterals (that fit none of the other classes), hexagons, and so forth.

Matching, identifying, exploring, and even making shapes with computer applications can be similarly motivating educational tasks. Drawing tools and other construction-oriented applications in computer environments can also provide symmetry functions. Playing "guess my rule" games, in which one person sorts shapes into dichotomous categories and others have to guess the rule for the sorting, can encourage close attention to the properties of geometric shapes. Such activities, along with discussions in which children consistently ask, "How do you know?" provide teachers with valuable assessment information on which to base future instruction.

Essential ←
Understanding 1*a*

Mathematical classification extends and refines everyday categorization by making more precise what we mean by "sides," "angles," "straightness," or other features that we attend to as we categorize mathematical objects.

➤ Essential
Understanding 3*a*

*Transformations can
be used to describe
differences between
an idealized image
of an object and
the way that it is
positioned in space
or seen by the eye.*

➤ Essential
Understanding 3*b*

*Under each
transformation,
certain properties are
invariant.*

Having students perform and discuss geometric motions has the potential not only to teach them about these transformations but also to improve their spatial skills, developing connections between mathematics and real-world observations, as Essential Understanding 3*a* suggests. Computers of all types, from laptops to tablets, with high-quality software, can be helpful, since the screen tools make motions more explicit. For example, computer environments can help children develop an understanding of congruence and symmetry as they specify the geometric motions of translation (slide), reflection (flip), and rotation (turn) to check whether two shapes are congruent by superposition.

The informal language "slide," "flip," and "turn" has some liabilities, by the way. Although "slide" might well be less ambiguous than "translation" for many children, who might associate the latter with *language* translation, "reflection" is a familiar word with exactly the right meaning. Further, "flip" and "turn" are often used rather interchangeably in common language ("turn the pancake over" means "flip the pancake"; and "flip the letter **R** upside down" could indicate either rotation or reflection), so why not use "reflection" and "rotation" as the natural language of instruction, even if one prefers "slide" in place of the more formal term "translation"?

Although such *rigid* motions maintain congruence, teachers know that not all transformations do (Essential Understanding 3*b*). A dilation, like a copy machine's reduction, maintains the "shape" of the object (more precisely, corresponding angles have the same measures, and the lengths of corresponding sides are in the same proportion) but not necessarily the size of the object (all sides might be, say, double the lengths of the original). Young children have considerable intuition about such scaling transformations, experience that may lead to interesting contrasts with the rigid motions and also provide another intuitive foundation for multiplication (arguably better and more accurate than "repeated addition"). Again, computer graphics programs have scaling functions that can complement other kinds of experience, adding important links among different representations. As children scale a figure, the pictured shape changes in size, but a report of measurements would show that the sizes of corresponding angles remain unchanged while the lengths of corresponding sides change multiplicatively, by the same factor.

The subject of angles deserves more attention. Angle and angle measure are important topics but typically are not quickly learned or easily taught. To understand angles, children must discriminate angles as critical parts of geometric figures, compare and match angles, and construct and mentally represent the idea of "turns," eventually integrating this with angle measure. For example, even preschoolers can and should differentiate between right angles and other angles—or else they cannot distinguish rectangles from non-rectangular parallelograms. Computer-based shape manipulation and navigation environments can help children mathematize these experiences. Especially important is understanding how turning one's body relates to turning shapes and turning along paths in navigation and learning to use numbers to quantify these situations with turns and angles.

The interaction between angles that we experience as we move and angles that we see on paper is also subtle. When we are traveling in a car and make a very sharp hairpin turn, for example, we experience the angle of the turn as "large," or "wide." However, the trace that we see—the "picture" of the angle on the map—is a very *small* angle, as illustrated in figure 3.2. In reality, two different angles are being measured: the angle at which the roads meet (a "sharp," acute angle) and the angle through which we change direction (a "wide," obtuse angle). Experience and attention, both with drawing and with moving, help children sort these out.

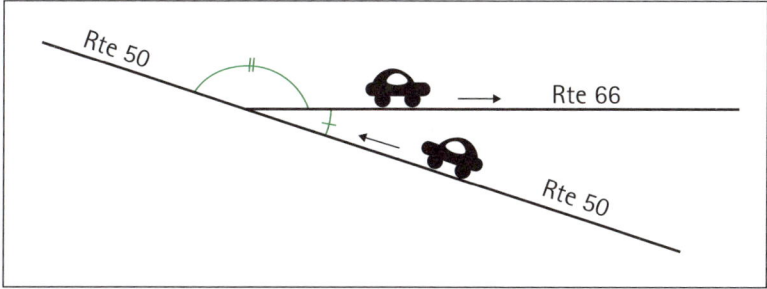

Fig. 3.2. A car traveling on route 50 makes a sharp right turn onto route 66; the angle appears on the map as the "small" marked angle, but the car turns through the much "larger" *supplementary angle* marked.

Across these topics, teachers who understand that classification schemes specify "for a space or the objects within it the properties that are relevant to particular goals and intentions" (Big Idea 1) can help children understand how shape classification "works" and how it differs across situations. The goal is to help children move beyond the visually based guessing that often underlies responses to such questions as "What shape is this?" Children realize that shapes are defined by their *relevant attributes,* and even that such relevance changes, depending on the context. As a simple example, in deciding which polygon is a rectangle and which is a non-square rhombus, orientation is *not* relevant. But in deciding which letter is a **b** and which is a **d**, orientation *is* relevant.

 Big Idea 1

A classification scheme specifies for a space or the objects within it the properties that are relevant to particular goals and intentions.

Offering Opportunities to Compose Shapes

Early childhood educators sometimes promote block building in preschool and kindergarten because they see such activity as "playing," which they prefer to "learning math" for children at these levels. Ironically, today's blocks were created in large part to teach mathematical concepts! Building with them *is* play *and* math learning! For a teacher who can see the rich mathematical opportunities that block play affords, building with blocks can become an effective way to help children develop geometric ideas, including composition of three-dimensional shapes, recognition of symmetries, and other spatial skills. In fact, preschool block building predicts students' mathematics achievement in high school (Wolfgang, Stannard, and Jones 2001).

Teachers who understand the mathematics of block building can pose questions that extend children's thinking without hijacking the play or turning fun into work. They might comment or ask a question about attributes of shapes (for example, "Which blocks stack well? Why?"), or composition of shape ("If you need more rectangular prisms, can you put together other blocks to make some?"), or symmetry ("Oh, your building looks completely symmetric... except for this part. Did you plan to do it that way?") Teachers can suggest new ideas or pose problems that allow them to join in or extend the play ("Are you going to build an arch over that?"). They should provide building opportunities and rich questions and comments to boys and girls equally because research shows that such activity helps both girls and boys develop important spatial skills (Casey et al. 2008; Kersh, Casey, and Young 2008).

Block building can also help children develop measurement concepts. Children have to struggle with length relationships in making a roof for a building. Length and equivalence are involved in substituting two shorter blocks for one long block, thus helping children develop ideas about measurement and shape composition.

Children can learn to compose and decompose two-dimensional shapes in many other productive ways. For example, they can make informal, creative pictures and tile with shape sets, such as pattern blocks (see fig. 3.3), and they can work with shape puzzles, including tangrams (see fig. 3.4). Puzzles that target a challenging but achievable level are generally intrinsically highly motivating and can be powerful contexts for learning. Eventually, such composition work can involve consciously making units of units. The work of decomposing, rearranging, and recomposing shape regions builds the notion that objects can be decomposed and composed in various ways, some of which may be more helpful in measuring their area.

Fig. 3.3. A child building with pattern blocks

Fig. 3.4. A child working with a tangram

Teachers who understand spatial structuring can help students form robust concepts of area. Once students can decompose a rectangular region into rows and columns, they will view the formula that they learn later, $A = l \times w$, as one way to compute the number of square units in that region. Combined with the decomposition previously discussed, this understanding of the area of a rectangle will allow them to determine the area of figures such as a "block L" (shown in fig. 3.5) by decomposing it into rectangles and determining the area of each and then combining the areas to find the total area of the shape.

Fig. 3.5. A block **L** is a shape that can be decomposed into rectangles.

Exploring Measurements of Space and Shapes

Big Idea 4

One way to analyze and describe geometric objects, relationships among them, or the spaces that they occupy is to quantify—measure or count—one or more of their attributes.

Teachers who understand measurement as embodied in Big Idea 4 can help children develop understandings and competencies that take them beyond the typical goal of building skill in using a conventional ruler. They can ensure that children understand the measurable attribute (such as length, area, and volume), the role of the unit of measure, the iteration of these units needed to measure, comparison of units and measures, and the composition of units (addition of units).

To develop these understandings, teachers often begin with the basic measurement question: How does this object compare with that one in a specified respect? For example, teachers may ask if the edge of a desktop is longer or shorter than a shoe, a hand, or a book—and not just any shoe, hand, or book, but a particular one. Once we determine that the desktop is longer than the length of a book, we might ask *how many* copies of the book (in this case, the unit of measure) does it take to match or make the same length? To make more comparisons, we move to a unit that we all can share—a *standard unit*: an inch, perhaps, using inch tiles or the like.

It is worth noting that this approach avoids the frequent but sometimes premature push to measure with varied units, because using multiple arbitrary units can be confusing. Measuring with familiar units such as inches invokes language that children hear and therefore is often more interesting and meaningful to them. Consistent use of these units may develop a model and a context for children's construction of the *idea of* and *need for* equal-length units, as well as the wider notion of what measurement is all about. Later, different units can be used to generate more explicit, generalized understanding of the role of standard equal-length units (such as centimeters or inches) and the inverse relationship between the size of the unit and the number of units needed to measure the length of an object.

In summary, teachers who grasp Big Idea 4 recognize three guidelines for teaching measurement to young children:

1. Measurement must be taught as more than a simple skill.

2. Children must be encouraged to solve real measurement problems, and, in doing so, to build and iterate units, as well as units of units.

3. Children must be guided to closely connect the use of manipulative units, rulers, and arithmetic on the number line.

Teachers who understand the foundational concepts of measurement can interpret children's understanding and ask questions that will lead them to construct these ideas. They realize that the principles of measurement are difficult for children and deserve more attention in school than they usually receive. They understand that the transition from informal to formal measurement needs much more time and care, with instruction in formal measure always returning to basic principles.

Using Learning Trajectories to Link Assessment and Instruction

Once the big ideas and essential understandings are clear, how do teachers choose, sequence, and assess instructional tasks? Learning trajectories can guide such choices. As students learn about a mathematical topic, they progress through increasingly sophisticated levels of thinking. At each level, children can solve a new type of problem. Further, the levels form a developmental progression, and certain educational experiences are particularly effective at each one. Thus, a learning trajectory has three parts: a learning *goal*, a *developmental progression* of levels of thinking, and associated *instructional activities* (Clements and Sarama 2009; Sarama and Clements 2009), as described below:

1. The goal is the students' attainment of the big ideas and essential understandings of the domain.

2. To achieve the goal, students progress through several levels of thinking (the developmental progression).

3. Students are aided in their progress by tasks and experiences (the instructional activities).

Figures 3.6 and 3.7 present parts of two learning trajectories for geometry and measurement adapted from Clements and Sarama (2009). In each trajectory, the goals are the big ideas and essential understandings from chapter 1. The left-hand column characterizes students according to the thinking that they are able to do at different levels in the developmental progression for the topic. For each level of thinking, this column also includes an example of student understanding and behavior. The right-hand column provides illustrations of the types of tasks that can help develop thinking at that level.

Thinkers at different levels in the developmental progression	Instructional tasks
Pre-area quantity recognizer Shows little specific concept of area. Draws mostly closed shapes and lines with no indication that they are intended to cover the specific region. 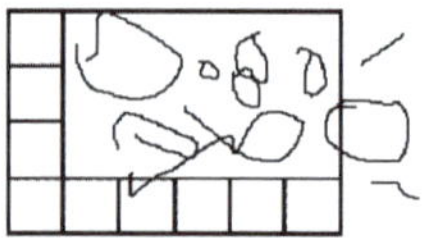 	Children intuitively compare, order, and build with many types of materials and increasingly learn vocabulary for covering a region of two-dimensional space.
Simple area comparer May compare areas of figures by using only one side or by estimating on the basis of length plus (not times) width. Asked, for example, which rectangular slice of bread is the "same amount" as a rectangle that is, say, 7 by 5 inches, one child chooses a 5-by-6 sample by matching the sides of the same length, and another child chooses a 6-by-6 sample, intuitively summing the side lengths.	Children determine which piece of paper will let them paint the biggest picture. They devise strategies such as cutting and rearranging to answer such questions.
Side-to-side area measurer Covers a rectangular space with physical tiles. However, cannot organize, coordinate, and structure a 2-D space without such perceptual support. In drawing (or imagining and pointing to count), can represent only certain aspects of that structure, such as approximately rectangular shapes next to one another. Covers a region with physical tiles, and counts them by removing them one by one.	Children's early attempts at measuring area might include tiling a region with a two-dimensional unit of their choosing, in the process discussing issues of leftover spaces, overlapping units, and precision. Discussions of these ideas lead children to partition a region mentally into subregions that can be counted.
May attempt to fill region, but leaves gaps and does not align drawn shapes (or aligns them only in one dimension). 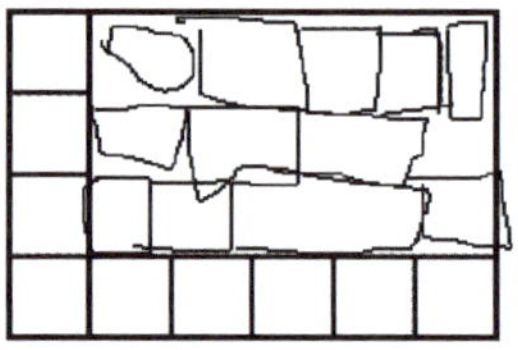 	

Fig. 3.6. A learning trajectory for spatial structuring and area measurement.
Based on Clements and Sarama (2009, pp. 177–80).

Thinkers at different levels in the developmental progression	Instructional tasks
Primitive coverer Draws a complete covering, but with some errors of alignment. Counts around the border and then, unsystematically, in the interior, counting some subregions twice and skipping others. 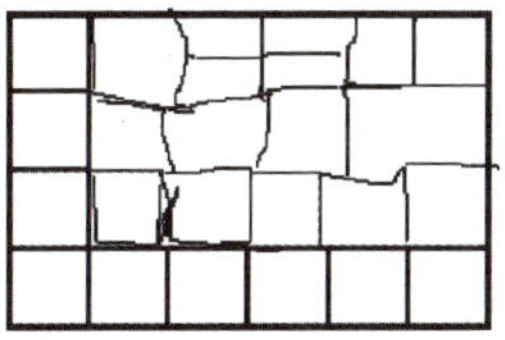	Children cover a rectangle by tiling with physical square tiles and then learn the drawing convention to represent two contiguous edges with a single line. They discuss how best to represent a tiling with no gaps.
Area unit relater and repeater Draws as shown for the primitive coverer. Counts correctly, aided by counting one row at a time and, often, by perceptual labeling.	Children discuss, learn, and practice systematic counting strategies for enumerating arrays.
Partial row structurer Draws and counts some, but not all, rows as rows. May make several rows and then revert to making individual squares, but aligns them in columns. Does not coordinate width and height. In measurement contexts, does not necessarily use the dimensions of the rectangle to constrain the unit size. 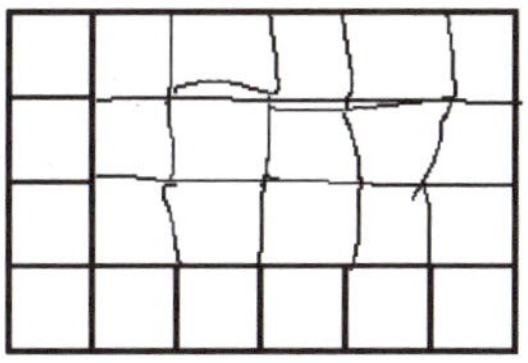	Children use squares of paper to measure areas to reinforce the use of the unit square as well as non-integer values. Children fill in ever greater numbers of missing sections and use descriptive language such as "bringing down" a row. 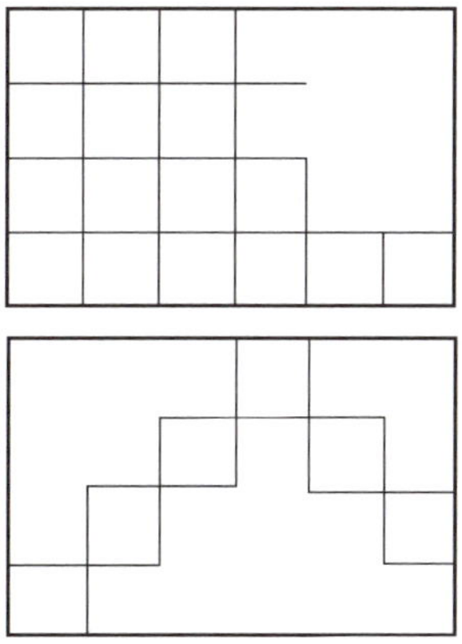

Fig. 3.6. *Continued*

Thinkers at different levels in the developmental progression	Instructional tasks
Row and column structurer Draws and counts rows as rows, drawing with parallel lines. Counts the number of squares by iterating the number in each row, using either physical objects or an estimate for the number of times to iterate. Those who count by ones usually do so with a systematic spatial strategy (for example, by row). If the task is to measure an unmarked rectangular region, measures one dimension to determine the size of the iterated squares, and eventually measures both dimensions to determine the number of rows needed in drawing. May not need to complete the drawing to determine the area by counting (as most younger children do) or computation (repeated addition or multiplication). 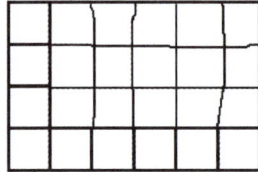	To progress, children need to move from local to global spatial structuring, coordinating their ideas and actions to see squares as parts of rows and columns. Children need to "fill in" open regions by mentally constructing a row, setting up a one-to-one correspondence with the indicated positions, and then repeating that row to fill the rectangular region. 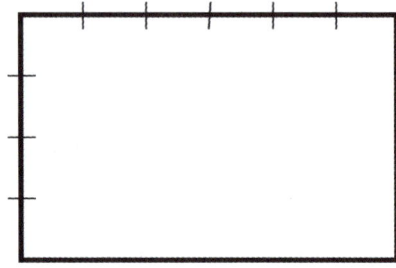
Array structurer Multiplicatively iterates squares in a row or column to determine the area of a rectangular region with linear measures or other similar indications of the two dimensions. Drawings are not necessary. In multiple contexts, children can compute the area from the length and width of rectangles *and* explain how that multiplication creates a measure of area.	Given two rectangles (and later, shapes made from several rectangles), children determine how much more space is in one than the other.

Fig. 3.6. *Continued*

Thinkers at different levels in the developmental progression	Instructional tasks
Pre-length quantity recognizer Does not identify length as attribute: "This is long. Everything straight is long. If it's not straight, it can't be long."	Children intuitively compare, order, and build with many types of materials and increasingly learn vocabulary for specific dimensions.
Length quantity recognizer Identifies length or distance as an attribute: "I'm tall, see?"	Teachers listen for and extend conversations about things that are "long," "tall," "high," and so forth.
Length direct comparer Physically aligns two objects to determine which is longer or if they are the same length. Stands up two sticks next to each other on a table, and says, "This one's bigger." 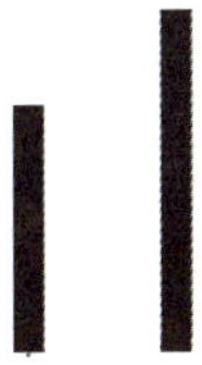	In many everyday situations, children compare heights and other lengths directly (who has the tallest tower, the longest clay snake, and so on). Children cut a ribbon the length of their arms and find things in the classroom that are the same length. Children order themselves (with teacher's assistance) by height in groups of 5 during transitions from one activity to the next during the school day.
Indirect length comparer Compares the length of two objects by representing them with a third object. Compares length of two objects with a piece of string.	Children solve everyday tasks that require indirect comparison, such as determining whether a doorway is wide enough for a table to go through. When asked to compare the length of two objects indirectly by use of a third object, children often *cover* the two objects by the third one, making indirect comparison impossible. Giving them a task such as comparing the lengths of two felt strips by use of a (wider) strip of paper can provide an opportunity to encourage them lay the third object next to each of the objects to be compared since if they cover the felt strips with the paper strip, they will have to make a guess on the basis of a visual comparison. If their guess is not correct, they can be asked how they could have used the paper to make a better comparison. Laying the paper strip next each of the felt strips can be modeled if necessary.

Fig. 3.7. A learning trajectory for length measurement.
Based on Clements and Sarama (2009, pp. 169–72).

Thinkers at different levels in the developmental progression	Instructional tasks
End-to-end length measurer Lays units end to end. May not recognize the need for equal-length units. The ability to apply resulting measures to comparison situations develops later in this level. Lays 9-inch cubes in a line beside a book to measure how long it is.	Children's thinking is stimulated by riddles such as, "You write with me, and I am the length of 7 cubes. What am I?" Children measure with physical or drawn units. Their thinking about length benefits from a focus on long, thin units, such as toothpicks cut to 1-inch sections. Explicit emphasis should be given to the *linear nature* of the unit. That is, children should learn that when they measure with, say, centimeter cubes, they should use the length of one edge that is the linear unit—not the area of a face or volume of the cube.
Length unit relater and repeater Measures by repeated use of a unit (but initially may not be precise in such iterations). Relates size and number of units explicitly (but may not appreciate the need for identical units in every situation). Relates size and number of units explicitly: "If you measure with centimeters instead of inches, you'll need more of them because each one is smaller." Can add up two lengths to obtain the length of a whole: "This is 5 long and this one is 3 long, so they are 8 long together." Iterates a single unit to measure. Recognizes, at least intuitively or in some situations, that different units will result in different measures and that units should be identical. Uses rulers with minimal guidance. Measures a book's length accurately with a ruler.	Children may be able to draw a line of a given length before they are able to measure objects accurately. Giving them line-drawing activities provides opportunities to emphasize starting at the 0 (the zero point, or *origin*) and to engage them in discussing how to measure an object by aligning it with that point as well as what the intervals and the numbers represent, connecting these to end-to-end length measuring with physical units. Children confront measurement with different units and discuss how many of each unit will fill a linear space. They make explicit statements to the effect that the longer the unit, the fewer "copies" of the unit are needed.

Fig. 3.7. *Continued*

Thinkers at different levels in the developmental progression	Instructional tasks
Length measurer Considers the length of a bent path as the sum of its parts (not the distance between the endpoints). Measures, understanding the need for identical units, relationships between different units, partitions of the unit, the zero point on rulers, and accumulations of distance: "I used a meter stick three times, then there was a little left over. So, I lined it up from 0 and found 14 centimeters. So, it's 3 meters 14 centimeters in all."	Children use a physical unit and a ruler to measure line segments and objects that require both an iteration and subdivision of the unit. In learning to subdivide units, they may fold a unit into halves, mark the fold as a half, and then repeat the process, to build fourths and eighths. Children create units of units, such as a "footstrip" made by tracing their footprint repeatedly on a strip of adding-machine tape. They then measure in different-sized units (for example, 15 footprints or 3 footstrips, each of which has five footprints) and accurately relate these units. They also discuss how to deal with leftover space—to count it as a whole unit or as part of a unit.
Conceptual ruler measurer Possesses an "internal" measurement tool. Mentally moves along an object, segmenting it, and counting the segments. Operates arithmetically on measures ("connected lengths"). Estimates with accuracy: "I imagine one meter stick after another along the edge of the room. That's how I estimated that the room's length is 9 meters."	Children use given measures to find the measures of figures. This is an excellent activity to conduct on the computer, using Logo's turtle graphics. 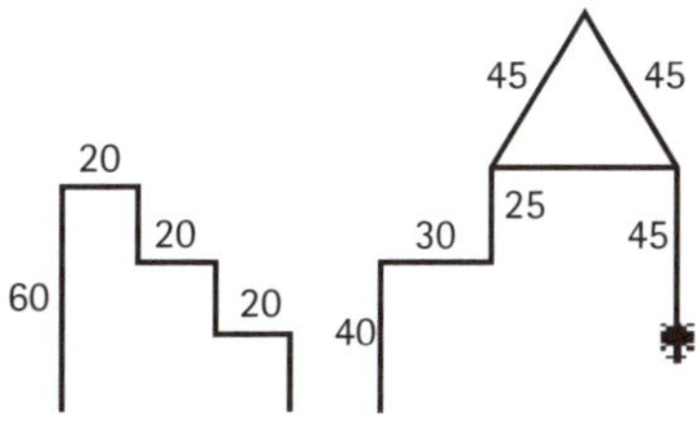 Children learn explicit strategies for estimating lengths, including developing benchmarks for units (such as an inch-long piece of gum) and composite units (for example, a 6-inch dollar bill) and mentally iterating those units.

Fig. 3.7. *Continued*

How do teachers use learning trajectories? The main way, of course, is to use a curriculum that has been constructed with deliberate attention to the trajectories of students' learning. However, any teacher can use them with any curriculum by modifying and enhancing the curriculum. That is, teachers can use formative assessment—ongoing monitoring of children's learning—to inform instruction.

Learning trajectories help such ongoing monitoring by providing behavioral indicators at each level of a given developmental progression and showing what *distinguishes* them from behaviors at levels just before and after, especially those levels that children are most likely to demonstrate in their classrooms (Clements and Sarama 2009; Sarama and Clements 2009). Teachers can use a variety of assessment strategies, including individually administered interviews based on developmental progressions, samples of children's work, and performance tasks that illuminate children's thinking. One of the most effective is assessment embedded in small-group curricular activities. That is, during teacher-led, small-group lessons, teachers can record the level of thinking demonstrated by each of four to six children in the group and use that awareness to modify the activity and plan for future lessons.

Such ongoing monitoring helps only if teachers use it to inform instruction. Learning trajectories assist by suggesting the *type* of activities that may be particularly beneficial for learners at each level in a developmental progression and how and why those activities may assist learning by building toward the next level of thinking. The activities listed in the figures and the original sources are simply illustrative of useful instructional tasks. Teachers who understand all three aspects of learning trajectories—the goal (the mathematics for students, and how it fits with the big ideas and essential understandings in chapter 1), the developmental progression, and these tasks—can modify their curriculum's activities to benefit children at different levels and enhance their instruction by substituting tasks as they see fit. To repeat, instructional tasks in learning trajectories are not the only way to guide children to achieve the levels of thinking embedded in the learning trajectories. They are intended only as examples of the types of instructional activity that help to promote thinking at the subsequent level. Thus, teachers can implement them, adapt them, or use them as templates to gauge the usefulness of other lessons, including those in their published curricula.

Conclusion

We have focused on the mathematical content of geometry and measurement and spatial thinking that is important to teachers of mathematics in prekindergarten–grade 2 to understand. In the process, we have also looked at the ways in which children think and learn related content, and the characteristics of tasks that can support both assessment and instruction. By using rich situations and tasks that probe children's knowledge of the full range of mathematical ideas, teachers can gauge their movement along developmental progressions. Teachers can then use this information to help children work steadily toward greater proficiency and more sophisticated mathematical thinking.

References

Barnett-Clarke, Carne, William Fisher, Rick Marks, and Sharon Ross. *Developing Essential Understanding of Rational Numbers for Teaching Mathematics in Grades 3–5.* Essential Understanding Series. Reston, Va.: National Council of Teachers of Mathematics, 2010.

Casey, Beth M., Nicole Andrews, Holly Schindler, Joanne E. Kersh, Alexandra Samper, and Juanita V. Copley. "The Development of Spatial Skills through Interventions Involving Block Building Activities." *Cognition and Instruction* 26 (July 2008): 269–309.

Clements, Douglas H., and Julie Sarama. *Learning and Teaching Early Math: The Learning Trajectories Approach.* New York: Routledge, 2009.

Common Core State Standards Initiative (CCSSI). *Common Core State Standards for Mathematics. Common Core State Standards (College- and Career-Readiness Standards and K–12 Standards in English Language Arts and Math).* Washington, D.C.: National Governors Association Center for Best Practices and the Council of Chief State School Officers, 2010. http://www.corestandards.org.

Dougherty, Barbara J., Alfinio Flores, Everett Louis, and Catherine Sophian. *Developing Essential Understanding of Number and Numeration for Teaching Mathematics in Prekindergarten–Grade 2.* Essential Understanding Series. Reston, Va.: National Council of Teachers of Mathematics, 2010.

Elia, Illiada, Athanasios Gagatsis, and Andreas Demetriou. "The Effects of Different Modes of Representation on the Solution of One-step Additive Problems." *Learning and Instruction* 17 (December 2007): 658–72.

Geary, David C., Mary K. Hoard, Lara Nugent, and Jennifer Byrd-Craven. "Development of Number Line Representations in Children with Mathematical Learning Disability." *Developmental Neuropsychology* 33, no. 3 (2008): 277–99.

Hofmeister, Alan M. "Elitism and Reform in School Mathematics." *Remedial and Special Education* 14 (November/December 1993): 8–13.

Kersh, Joanne, Beth M. Casey, and Jessica Mercer Young. "Research on Spatial Skills and Block Building in Girls and Boys: The Relationship to Later Mathematics Learning." In *Contemporary Perspectives on Mathematics in Early Childhood Education*, edited by Olivia N. Saracho and Bernard Spodek, pp. 233–51. Charlotte, N.C.: Information Age, 2008.

Lobato, Joanne, and Amy B. Ellis. *Developing Essential Understanding of Ratios, Proportions, and Proportional Reasoning for Teaching Mathematics in Grades 6–8.* Essential Understanding Series. Reston, Va.: National Council of Teachers of Mathematics, 2010.

Lyons, Michel, and Robert Lyons. *Meta-Forms: Logic Builder.* Montreal, Canada: FoxMind Games, 2007.

National Council of Teachers of Mathematics (NCTM). *Principles and Standards for School Mathematics.* Reston, Va.: NCTM, 2000.

———. *Curriculum Focal Points for Prekindergarten through Grade 8 Mathematics: A Quest for Coherence.* Reston, Va.: NCTM, 2006.

———. *Focus in High School Mathematics: Reasoning and Sense Making.* Reston, Va.: NCTM, 2009.

Otto, Albert Dean, Janet H. Caldwell, Cheryl Ann Lubinski, and Sarah Wallus Hancock. *Developing Essential Understanding of Multiplication and Division for Teaching Mathematics in Grades 3–5.* Essential Understanding Series. Reston, Va.: National Council of Teachers of Mathematics, 2011.

Ramani, Geetha B., and Robert S. Siegler. "Promoting Broad and Stable Improvements in Low-Income Children's Numerical Knowledge through Playing Number Board Games." *Child Development* 79 (March/April 2008): 375–94.

Rodriguez, Diane, Rene S. Parmar, and Barbara R. Signer. "Fourth-Grade Culturally and Linguistically Diverse Exceptional Students' Concepts of Number Line." *Exceptional Children* 67 (2001): 199–210.

Sarama, Julie, and Douglas H. Clements. *Early Childhood Mathematics Education Research: Learning Trajectories for Young Children.* New York: Routledge, 2009.

Smith, Linda B., Susan S. Jones, Barbara Landau, Lisa Gershkoff-Stowe, and Larissa Samuelson. "Object Name Learning Provides On-the-Job Training for Attention." *Psychological Science* 13 (January 2002): 13–19.

Vanbinst, Kiran, Pol Ghesquiere, and Bert De Smedt. "Numerical Magnitude Representations and Individual Differences in Children's Arithmetic Strategy Use." *Mind, Brain, and Education* 6 (September 2012): 129–36.

Wolfgang, Charles H., Laura L. Stannard, and Ithel Jones. "Block Play Performance among Preschoolers as a Predictor of Later School Achievement in Mathematics." *Journal of Research in Childhood Education* 15 (Spring/Summer 2001): 173–80.

Titles in the Essential Understandings Series

The Essential Understanding Series gives teachers the deep understanding that they need to teach challenging topics in mathematics. Students encounter such topics across the pre-K–grade 12 curriculum, and teachers who understand the big ideas related to each topic can give maximum support as students develop their own understanding and make vital connections.

Forthcoming:
Geometry and Measurement for Teaching Mathematics in Grades 3–5

Visit www.nctm.org/catalog for details and ordering information.